LIGHT

Inquiry and Insights

An inquiry-based course in optics

Third Edition

David Gray Schuster

Western Michigan University

A text for both conceptual and algebra-based courses

Originally developed at Western Michigan University
for the Physics 1800 course for prospective teachers

Kendall Hunt
publishing company

Cover image © JupiterImages Corporation

www.kendallhunt.com

Send all inquiries to:

4050 Westmark Drive

Dubuque, IA 52004-1840

Copyright © 2005, 2007, 2010 by David Gray Schuster

Revised Printing 2008

ISBN 978-0-7575-7299-9

Printed in the United States of America

10 9 8 7 6 5 4 3 2

Various faces of science

This book emphasizes three equally important faces of science, viz. science as process, science as product, and science as application.

Science as *process* refers to the inquiry processes by which scientific knowledge about nature is developed, i.e. it is about 'doing science' – exploring phenomena, posing questions, seeking patterns, developing concepts, laws and theories. The book approaches each new topic in this spirit, modeling these processes to develop the subject matter. This face of science could be called 'science-in-the-making'.

Science as *product* refers to the current state of scientific knowledge; a major intellectual achievement that scientists have produced over the centuries, and in terms of which we understand the way nature works. This second face could be called 'already-made-science'. The book organizes this knowledge in a way which reflects the structure of the discipline, identifying the underlying 'powerful ideas' of each topic. This book thus presents an organized coherent structure of concepts, principles, and models, and is not a set of fragmented science activities.

Science as *application* has two meanings. Firstly, the application of our scientific knowledge to explain a range of phenomena, make predictions, and solve problems. Secondly, the application of science to technology, e.g. new devices, instruments, inventions, etc. The book has a particularly strong emphasis on the problem-solving aspect, and in fact uses problems as a teaching and learning vehicle to further develop understanding of the science. One does not truly understand something until one can *use* it effectively in various situations! There is also some application to real world situations, e.g. the seasons, and to technology, e.g. to optical instruments, and of course application to everyday situations, but in general the course focus is on the science rather than on technology. This third face, or application aspect of science, might be called 'science power', or perhaps 'idea power'

The three faces are different but complementary, and the book emphasizes them all. Note that conventional courses and textbooks often tend to neglect the first (process) face, while by contrast some inquiry-based courses tend to concentrate primarily on it at the expense of the others. All are important for a full understanding and appreciation of the human endeavor of science.

Various languages of science - multiple representations

Besides having various *faces*, as described above, science also has various *languages*. Scientific ideas – concepts, principles, laws, procedures, models, theories etc – can be conceptualized, described and communicated in various ways; in words, in diagrams, in graphs, in charts, in mathematical equations, etc. We can think of these as various *languages* of science – ways to represent and express the ideas. These multiple representations of scientific ideas are of course *complementary* to each other.

All of these representations are used in this book, and the learner should aim to become proficient in all of them, be able to translate between them, and to draw on the language most useful for the purpose at hand.

The cognitive aspects of understanding physics

In the book we talk not just about physics but about how to think in learning and understanding physics and solving problems. That is, we discuss cognition as well as content. Thus we discuss significant modes of reasoning in physics, namely principle-based reasoning, case-based reasoning and experiential-intuitive reasoning. This is not simply in general terms; for example in each chapter we consider "topic knowledge subassemblies' relevant to the particular topic at hand. We also encourage metacognitive reflection on the topic and one's own learning, and include an appreciation of the epistemological aspects of science.

Some structural features of the book

The course comprises geometrical optics, i.e. the geometrical or 'ray' optics aspect of light. There are also two special chapters, the first on the nature of science and learning science, the other on the basic mathematics of similar triangles.

By design, the approach to each topic reflects the nature of science, two aspects in particular – the processes of scientific inquiry and the goal of science of explaining a range of situations in terms or relatively few fundamental concepts and principles, or 'powerful ideas'. These aspects occur implicitly through the approach to each topic, but are also discussed explicitly along with content.

Mathematical aspects are for the most part introduced as needed in the context of the physics topics at hand. The geometry of similar triangles and side ratios is an aspect common to many topics in geometrical optics and hence a separate short mathematical chapter is devoted to developing these ideas, ahead of the chapters on shadows, apertures, reflection and refraction.

Note that this is essentially a *physics* course, but within each topic we take opportunities to reflect on the learning issues involved, and sometimes on the implications for teaching the topic. Thus where appropriate our discussions complement physics content with educational aspects.

Flexibility: conceptual and mathematical treatments

The book is flexible that it can be used for both purely conceptual courses and algebra-based courses. Each topic is approached in a qualitative conceptual manner first, to develop conceptual understanding and physical insight. This is then complemented by a more mathematical treatment. The math for the most part involves similar triangles, ratios, and basic algebra. Very little prior mathematics background is assumed, since most of mathematical ideas are introduced and explained in the context of the physics topic at hand. A close correspondence is always made between the mathematical representation and the physical situation. An intermediary 'verbal logic' approach is sometimes used to solve quantitative problems, as a bridge connecting conceptual understanding and formulas.

This topic structure means that conceptual courses can simply omit the more mathematical sections of a chapter, without losing the coherence or integrity of topic development, while algebra-based courses can include the mathematics as well. Another option is to treat some topics conceptually and others both conceptually and mathematically, as the instructor prefers and as time allows.

This conceptual/mathematical design structure has two advantages. Firstly, for learning science it is desirable in any case to develop conceptual understanding and physical insight before formalism, equations and quantitative calculation. Secondly, it means the book can be used flexibly for either a purely conceptual or an algebra-based course (or a judicious mixture) by appropriate choice of sections. Correspondingly, the problems for each chapter are likewise grouped so that they can be readily selected for either type of course.

The extent of the course

The course is designed for half a semester. To shorten the course, certain chapters can be omitted if desired without loss of coherence. For example one could omit the chapter on the variation of light intensity with distance and angle, and its applications to climate and seasons. In addition, the mathematical aspects of various sections may be omitted if a purely conceptual course is desired. Thus for example, if one for some reason did not wish to teach the basic mathematics of shadows, then one could include only the conceptual and qualitative sections, and solve problems using scale ray diagrams and string simulations. Similarly the sections on apertures, reflection and refraction include mathematical sections which could be omitted. Nevertheless, we feel that such physics topics are in fact the ideal context in which to teach the relevant mathematical ideas! An understanding of the basic mathematical ideas, and their connection to physical situations, is important and valuable well beyond these particular topics, so it would be a pity to omit them. In fact we view the course as teaching mathematical and thinking skills along with science content! However, omitting some of the mathematical aspects is certainly a way to shorten the course, or to make it purely conceptual without any formal mathematics.

Questions and problems

Sets of questions and problems are provided at the end of each chapter. Many of them are structured problems, comprising a problem situation followed by a number of sequenced sub questions, to address multiple aspects of understanding and physics thinking. Thus a structured problem may include, where appropriate, conceptual questions, qualitative questions, constructions, diagrams, algebraic questions, dependencies, quantitative calculations, etc. Problems do not simply require an answer to a single question posed on a selected aspect of a situation, but an understanding of the physics of the whole situation. This contrasts with some of the end-of-chapter problems found in some texts, which tend to be predominantly routine exercises or formula-based quantitative calculations Some of the problems are targeted at specific aspects of a chapter or topic, and will be useful in teaching those aspects, while others are 'combination' problems which require students to understand the whole chapter.

Structured problems of this nature may be considered as 'teaching and learning' problems. By working through them, learners enhance their understanding of a topic and develop their ability to apply the concepts and principles to new situations. The structured nature of the problems has two features; it ensures that learners engage with a full range of aspects of the situation, and it can act as scaffolding in difficult problems that they might not otherwise be able to start. Of course the scaffolding aspect can be absent for other problems once learners have become proficient.

The problems thus serve well as formative assessment. They may be used for class examples, homework, or independent study and practice.

Note that in the concept development phase of a topic, learners engage actively in generating the science, (science-in-the-making), while likewise in the application phase, they engage actively in problem solving, (science-at-work). The active engagement aspect in each of these is important for learning physics.

The problems, although they occur at the end of chapters, are a particularly important part of the course. They provide formative 'assessment for learning'. The problem sections of the book are distinguished by having a line border around each page.

Assessment

Sets of further questions and problems, of a similar nature, have been produced for each topic. These can be used for both formative and summative assessment in quizzes, tests and examinations. They are available but not yet published.

Roles of textbook and student notes

This is intended to be a 'working text' in that here is space on each page for student notes, ideas, responses, diagrams etc. In many cases, concepts and principles are developed in a guided inquiry fashion by students during the course, and these are then written in space allowed for it in the text, and book pages also have large margins for student notes and responses. The book is not however a set of worksheets; students also have their own course notes in which to keep a full record of the course, including discussions, group work, lab activities, experiments, data, problems & solutions, assignments, reflections, etc.

Acknowledgements

The course owes a debt to the insights of those who developed and implemented the original SCI 180 course at Western Michigan University, in particular Larry Opliger, Bob Poel and Bill Merrow. Various aspects of the current approach draw on ideas from the ISLE project of Alan van Heuvelen and Eugenia Etkina, PSSC Physics, Powerful Ideas in Physical Science and Constructing Physical Understanding.

I would like to thank Adriana Undreiu, Eric Arsznov, Betty Adams and Fang Huang, who as graduate instructors worked with me on the course 'teaching team' and contributed valuable ideas to both the overall design and matters of detail. Not to forget all the students who took the course while it was still 'under construction', and who had to deal with ongoing 'roadworks' each week, using draft materials that came out 'just in time'. We all thank our students, who keep teaching us things.

The later part of the project was supported in part by the Faculty Research and Creative Activities Support Fund of Western Michigan University, the Michigan Space Grant Consortium, and the National Science Foundation under grant n DUE-0536536. Any views expressed in this material are those of the author and do not necessarily reflect the views of the grant agencies.

David Schuster

CONTENTS

Chapter 1

SCIENCE AND SCIENCE EDUCATION

This chapter introduces important ideas about science and learning science. These are further developed throughout the book and are reflected in the approach to each physics topic.

It seems useful to introduce these overarching ideas in the first chapter, but note that you are unlikely to fully appreciate them until you have experienced particular cases in the topic chapters to come. Then you will have specific examples to help make sense of the general ideas. A suggestion is thus to read through this first chapter to get an overview of what it is all about, then study the physics in the next few chapters, and after that return to this chapter again. Note that it is important not to skip the chapter, since it describes the approach we take to science and science learning, and explains the rationale behind it.

SECTIONS

1-1 WHAT IS SCIENCE?

A question

Let us pose an apparently simple question: *What is science?*

The single word "science" surely carries a heavy burden – can one word possibly convey multiple facets of science?

It may help us think if we expand the question a bit:

> *What does science involve? What is it all about? What aspects are there? What do scientists do?*

So – what comes to mind when you think of science? Everyone will already have ideas about this, so let's start with an activity to gather these ideas.

Introductory activity: ideas about science

Suggest as many aspects of science as you can. Basically, write down everything that comes to mind when you think of 'science'.

List your own ideas

Add more ideas from the group

Add further ideas from instructor

Next we will gather everyone's ideas, cluster them according to type, and see what emerges.

Grouping these ideas – cluster diagrams

Draw a consolidated diagram of the various aspects of science that came out of peoples' suggestions, clustered into groups according to type.

Cluster diagram

Then discuss the nature of each cluster, and see if this sheds light on the question "what is science?"

The next section describes three broad faces of science, some of which you may have thought of in the exercise above.

1-2 FACES OF SCIENCE

When people speak of 'science', some rather different views of science emerge, which may seem to be at odds with one another.

Science in fact has several faces, and here are three important ones: Firstly, a face that is seeking to know – call this "science-in-the-making". Secondly, a face that already knows – call this "ready-made-science". And thirdly, a face that puts science knowledge to use – call this "science-at-work".

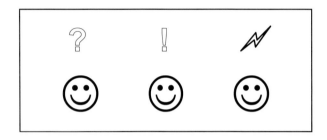

These three faces view the natural world in distinctively different ways.

Science-in-the-making is not a face very familiar to most people – yet it is the science that scientists actually *do*. In this view, science is a *process*, of creating new knowledge. This science is active and continually progressing. This face – how science is done – should play a large role in the school curriculum, but has often been neglected.

Already-made-science is the face of science which is most familiar to people. It is the science we read about in textbooks and magazines, the science we watch on TV, and the science that comprises much of the instruction in school. In this view, science is an existing *product* or body of knowledge. It is made up of known, facts, formulas, definitions, concepts, laws and theories.

Science-at-work is a face interested in putting science to use – applying it to explain, solve problems, and predict, as well as to invent, design and build devices.

Science curricula and textbooks have often been unbalanced in regard to these three faces, concentrating predominantly on the results of science. School science has been very much a presentation of already-made knowledge, providing students with established facts, formulas, definitions and laws to be learned. This science tends to be presented as a 'rhetoric of conclusions', as Schwab termed it.

By contrast, until fairly recently there have been relatively few curricula emphasizing the discovery face of science. This involves approaching science as inquiry and modeling the processes by which science develops. This book adopts an explicit processes-of-science curriculum structure to do so, approaching each major topic as in inquiry-based fashion (science-in-the-making). Once the concepts and principles are developed, the course emphasizes a deep understanding of accepted science (made-science), and the ability to apply this to new situations and problems (science-at-work). Thus all three faces are present in a balanced way.

We next look in more detail at each of the faces.

1. Science-In-the-Making

  This face of science, science-in-the-making, offers a perspective on what it is like to actually 'do' science. Rather than viewing the natural world from hindsight, it takes an inquiring forward-looking attitude, characteristic of the working scientist.

This face sees scientists struggling with questions about nature which have no easy answers. For that matter, they don't even come packaged with methods for investigation; these have to be devised. Difficulties are not minor obstacles but the very 'stuff' of doing science. Science becomes very much a problem-solving challenge, a process of generating new knowledge. Scientists are imaginative and inventive in tackling the questions at hand. They are observers, experimenters and decision-makers rather than rule-followers. They decide what aspects of the world to observe, how to design their explorations, what meaning to give to their observations, and how to interpret them in terms of concepts and theories they invent and refine.

We also see different scientists proposing different, competing models and theories to explain the same natural phenomena. Examples would be the different theories of heat, of light and color, or the development of the particle and wave theories of light. Each can explain some of the observed behavior, but may fail on other aspects, and ultimately some theories may be discarded, superseded or refined. And finally, we see scientists actively engaged in critiquing other models and persuading others to their point of view, for you see, the facts do not speak for themselves. The heart of "doing" science lies in the interpretation or making sense of the facts, by inventing concepts, models and theories, based on evidence. This process is inductive to a considerable extent, though not entirely so. Science-in-the-making is a creative intellectual endeavor. We might also term this face 'science as *process*'.

Notes

2. Ready-Made-Science

Most of our experience with science is of the ready-made kind. (We might also call it 'already-made'). Here, insights about the natural world are *provided to us* in the form of facts, formulas, laws and theories. That is, a body of knowledge. For example, Newton's laws of motion or the theory of evolution. Acquiring already-made-science is like picking apples from a tree; the fruits of scientific discovery are there to be plucked. Science in this sense is seen as a 'body of knowledge' , laid out for us based on the earlier work of others. Thereafter, working with the knowledge will be largely deductive. We might also term this face 'science as *product.*' The body of knowledge produced by many scientists over the centuries is a remarkable human achievement, and is now available 'ready-made' if we wish.

Find an example of ready–made textbook science, and insert or append

Of course just because the end-product results may be provided to us does not mean that the science will be easy to understand – it was challenging to come by originally, and will likely be demanding to understand and use.

Scientific facts, concepts, laws, and theories are useful because they enable us to make sense of the complexities of the natural world. To most of us, ready-made-science is clearly *useful.* But more than that it may have the appearance of being quite *certain.* It seems to have a truth-like quality. Thus we unquestioningly learn the facts and use the laws. But we rarely take a hard look at the *processes* through which those insights were developed. Ready-made-science looks at the products of scientific research with *hindsight,* and this often hides any sense of what it is really like to "do" science. The idea that scientific knowledge is *invented*, is based on evidence, and may be improved upon in the future, is largely absent for ready-made-science. And even if these things are discussed, the processes responsible for producing the insights are often reduced to systematic steps of a so-called 'scientific method', while the problems scientists encountered along the way are not discussed or are portrayed as minor obstacles to the attainment of truth.

Notes

3. Science-at-work

 To the first two faces we can add a third: a face which seeks to *use* the science. Use it to *explain, predict*, and *solve problems*.

 Another aspect of this face seeks to put the science to practical use, e.g. in designing, constructing and inventing things; this 'applied' face is familiar in technology and engineering.

We can call this face *science-at-work,* or alternatively 'science as *power'*.

Note that this face is also important for *learning* science – often it is only when one has to apply the science to solve problems that one really comes to understand it fully. If one can state knowledge but not apply it, one's knowledge is said to be 'inert'

So – what is science?

An answer to the original question "what is science?" should involve all three faces, and possibly other aspects too. Correspondingly, science instruction should be designed to reflect all three faces, otherwise learners will get a distorted and incomplete sense of science. Therefore, to the extent possible, students should themselves explore phenomena, produce knowledge, and apply it. In this way, whether they themselves will become scientists or not, they can develop a balanced perspective on what science involves, a knowledge of its important results, and the ability to use them in thinking scientifically. Science is a way of thinking and acting as well as a store of knowledge.

We might also name these three faces *Process, Product* and *Power.*

Notes

Acknowledgement. The idea of two faces of science occurs in *Science in Action* by Bruno Latour, Harvard University Press 1987, and we have modified and extended it to three faces. We also drew on an earlier version of two faces by Robert Hafner, David Rudge and Bob Poel at Western Michigan University.

1-3 SCIENCE EDUCATION AND REFORM

Science education approaches

When people talk of 'science education', what notions of science do they have in mind, and what notions of education? Are they thinking of science as process, science as product, science as application, or all of these? How does this affect their view of what science education should include? Are they thinking of education in terms of curriculum, content, pedagogy, learning, teaching, materials, assessment, or any or all of the multiple aspects of education?

Science education should ideally be about 'science-in-the-making' as much as about 'already-made-science' yet the latter has tended to dominate in conventional instruction and textbooks, where science is often presented as known facts, formulas, laws and theories. And the 'application' or problem-solving aspect has been less than ideal, often consisting of routine formula-based exercises requiring little conceptual understanding.

Three aspects of reform in science education which have gained prominence in recent years are *inquiry-based* approaches to content, problems requiring *conceptual understanding*, and understanding of the *nature of science*.

In the United States in K-12 grades there is now a national commitment to teaching science *as inquiry*. Inquiry refers to pedagogy that reflects the investigative approach used by scientists to discover and construct new knowledge. That is, inquiry teaching reflects the *processes* of science as well as the content. The processes of science include observation, questioning, exploring, experimental investigation, development of concepts and laws, and testing them, as well as subsequent application to problems and new situations.

At college level, reports by the US Department of Labor, the National Science Foundation, the American Institute of Physics and the ABET engineering accreditation organization suggest that graduates should know how to learn, be able to use the processes of scientific inquiry, be able to use expert-like approaches to solve problems and be able to design investigations.

It is well documented that students emerge from conventional science courses and having passed fact and formula-based tests, with misconceptions about both the subject-matter and what science is all about. An extensive bibliography by Pfundt and Duit covers research on students' conceptions in many areas of science.

Science Education Reform and Science Curricula

Calls for educational reform are not new. Problems with science instruction have long been recognized, yet conventional fact-based educational approaches still dominate. Etkina and van Heuvelen note that as long ago as the 17th century John Amos Comentus asked: 'How many of those who undertake to educate the young appreciate the necessity of first teaching them how to acquire knowledge?' John Dewey responded to this rhetorical question 300 years later by saying that science education has failed because it "has been so frequently presented just as so much ready-made knowledge, so much subject-matter of fact and law, rather than as the effective method of inquiry into any subject-matter".

There are indeed various physics courses, materials and pedagogical approaches that address these issues, often with reasonable success. Yet for the most part conventional practices persist. Despite cogent arguments in their favor the reforms still do not have the impact or acceptance hoped for.

We suggest that a problem might be that the ideas and approaches are not built structurally into curriculum and materials design, as an *overt* theme or organizing principle. Without this, content tends to dominate by default, despite intentions. Our hypothesis then, is that an explicit structure is required, supported by matching materials. This is the rationale for the design of this physics course. This and the ISLE project at Rutgers University are among the curricula we know of which use an *overt* scientific-inquiry framework.

A scientific inquiry approach to topics

Our course takes a predominantly *inquiry-based* approach to science. Thus we will "do science for ourselves" wherever possible. This means that we will investigate phenomena and construct many of the scientific concepts and principles involved, rather than simply being 'presented' with known scientific facts and formulas. Too often, science has been taught as a 'rhetoric of conclusions' (Schwab), but this will not be true for this course. We will approach science using observations, investigation, evidence and reasoning.

Process, product and power. As we have seen, science has three different but complementary faces, which we may call science as *process*, science as *product*, and science as *power*. The first looks to the *development* of science, the second to the *results* of science, and the third to the *application* of science. All are important, and to properly appreciate science we need to experience all of them. Hence they are built explicitly into our course. This approach also accords with national standards and guidelines for inquiry-based science education.

In the past, many curricula reflected a 'direct instruction' approach, where the accepted science was simply presented to students. This reflects an unspoken assumption that knowledge can simply be transmitted and absorbed. A quote from Paul Tillich serves as a comment: "The fundamental pedagogical error is to toss answers, like stones, at the heads of those who have not yet asked the questions"

By way of reaction, a swing toward 'inquiry-based' instruction began restoring the first face to its rightful place – but sometimes at the expense of the second and third faces! The pendulum can swing too far the other way. Thus all too commonly we see 'inquiry' instruction where students have to try to 'discover' everything themselves, via largely unguided 'activities'. Since it is well nigh impossible for students to construct the whole coherent framework of an area of science in this way – after all it took scientists centuries – this extreme approach tends to be inefficient, fragmented and often frustrating. It does not represent science authentically, takes an awful lot of time, and eats away into time for

9

understanding the accepted science and learning to apply it in various situations. It is true that such 'activities' may well involve various 'process skills' of science, such as measuring, taking data, tabulating, classifying, graphing, using equipment, etc., but this alone does not make them science!

Some other misunderstandings of 'inquiry' are as follows: that it is synonymous with 'hands-on" or synonymous with eliciting students' 'prior knowledge'; that the teacher asks but does not answer questions; that the process is important not the outcome, and that closure of an inquiry lesson around the accepted science is to be avoided. These are misinterpretations of inquiry in instruction.

Our course takes a balanced approach to inquiry, with respect to both science and pedagogy. It makes use of *learning cycles* to help achieve this, as described in the next section.

Notes

1-4 SCIENCE INSTRUCTION AND LEARNING CYCLES

Translating faces of science into instructional design

We have discussed three important faces of real science. The challenge for science education is: how can we best reflect these faces in instruction? How do we approach a science topic? What should students experience? Broadly, how could we design curriculum, teaching and textbooks to reflect the nature of science in a balanced way? Ideas? What would the approach to a particular science topic of your choice look like? Think and discuss. To make the ideas concrete, try sketching out a rough teaching plan for that topic.

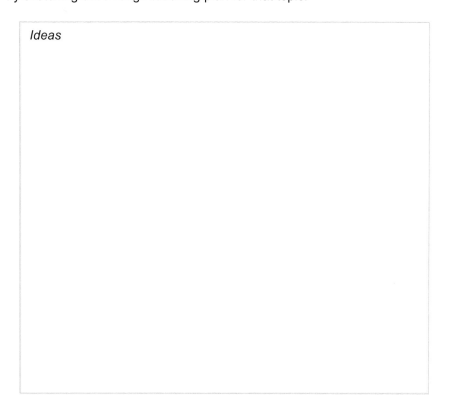

Ideas

LEARNING CYCLES

Even where teachers are aware of the characteristics of both science and student learning, they still face the task of transforming topic subject-matter into an instructional sequence that reflects these aspects. To assist in instructional design, 'learning cycles' have been devised. A learning cycle is a framework for teaching and learning science that explicitly reflects the important aspects of real science, and provides an instructional sequence for this. Thus a learning cycle will take a scientific inquiry approach to each topic as well as reflecting good pedagogical strategies.

Karplus and Atkin proposed the first science learning cycle in 1962 and variations and extensions have followed, such as the '5–E' cycle. These are described below

A. The Karplus Learning Cycle

The original learning cycle proposed by Karplus and Atkin in 1962 has three major phases, as shown in the table and figure. (Note that variations may occur in the names given to phases).

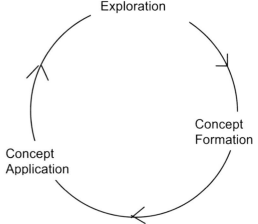

PHASE	DESCRIPTION
"Exploration"	Observing the phenomenon, exploring its behavior, and investigating systematically.
"Concept formation"	Guided invention/introduction of relevant scientific concepts and formulation of laws, principles and/or models for the phenomenon.
"Concept Application"	Application of these concepts, principles and models to solve problems, explain and predict.

You will recognize that this learning cycle incorporates the three important 'faces' of science we discussed, viz. the *making* of science, the *results* of science and the *application* of science.

The exploration and concept formation phases together reflect a *scientific inquiry* approach to science. These phases are inquiry-based and fairly *inductive* in character (though not purely so).

The application phase is mostly *deductive* in nature. Here students develop the ability to *apply* the science and solve problems.

Note also that in having to apply the science knowledge, students are further enhancing their understanding of it. In fact many problems are specially designed with such active learning in mind. Such 'teaching and learning problems' can used as teaching examples and as formative assessment ('assessment for learning').

Note that one may traverse the learning cycle a number of times in developing various aspects of a topic. Furthermore the phases are not sharply separate but overlap, and one may also link back and forth between them, since each phase can act to inform the others. Thus while the learning cycle is a very useful way to conceptualize science and instruction, it is not a rigid prescription.

We shall not cease from exploration,
and the end of all our exploring
will be to arrive where we started
and know the place for the first time.

                                      – T S Eliot*

B. The "5-E" Learning Cycle

The original three-stage cycle is sometimes expanded into five stages, as in the so-called "5-E" learning cycle. It contains the same three basic phases as the Karplus cycle, but adds an "engagement" phase at the beginning, and an "evaluation" phase at the end. The phases are all named to start with "E", simply for ease of recall.   Again, one finds some variation in the names given to the phases

	Phase	Description
E1	*Engage*	*Engaging* learners' interest, curiosity and wonder about the topic, and raising relevant questions. Also called 'Invitation to learn'.
E2	*Explore*	*Exploring* the nature & behavior of the system and investigating systematically.
E3	*Explain*	*Explaining* the phenomenon through the development of relevant concepts, laws, principles and/or models. Called *'concept formation'* in the Karplus cycle.
E4	*Employ*	*Employing* these scientific ideas to solve problems, explain and predict. Often called Extend or Elaborate. Called *'concept application'* in the Karplus cycle.
E5	*Evaluate*	*Evaluating* by reflecting back and identifying the essence of the topic, what it is all about, what has been learned, and how. Better called *Reflection*.

Notes

Note that the 5-E cycle incorporates the three Karplus phases: they are now called Explore, Explain and Employ, but the ideas and intent remain the same.

Note that the 'Employ' phase is more often called 'Extend', or sometimes 'Elaborate'. The Karplus phrase 'Concept Application' is probably clearer.

There may be some ambiguity about the term 'Explain'. Here it means formulating scientific concepts and principles that would serve to explain our exploration data. The Karplus phrase 'Concept Formation' may be clearer.

There may also be ambiguity about the term 'Evaluate'. Here it means 'reflect on' – i.e. consciously thinking back on the topic, picking out its essence and purpose, and how best to understand it. It does not mean evaluate in the sense of assessment via a quiz. Thus 'Reflect' may be a better term, but it does not start with an 'E'!

At the end of a topic, after the learning cycle, we can imagine adding a possible "Expand" phase; here we would go beyond the science conceptual development already completed, and expand our perspective, showing how the topic relates to other areas of science, or show important applications in technology, engineering or medicine. Ending this way can also complement the initial 'engage' phase. However this extra phase is not commonly used.

The 5-E cycle diagram

The figure on the right shows one way of representing the five "E" phases diagrammatically.

Note that some 5–E diagrams show the Evaluate (Reflect) aspect in the *center*, connected to all the others, because reflecting about what's going on can occur at any time during science learning.

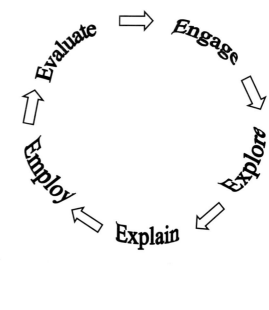

Comparison of Karplus and 5-E learning cycles

Since the Karplus and 5-E cycles are related it is useful to compare them directly as in the table below.

Karplus cycle		5-E cycle
–	1	Engage
Exploration	2	Explore
Concept formation	3	Explain
Concept application	4	Employ/Extend
–	5	Evaluate (Reflect)

Both cycles reflect three important aspects of science learning, though they may name some of them differently. The 5-E system adds an explicit engagement phase and reflection phase.

Example of five phases for a particular topic

It may be useful to have an example of what the phases might look like for teaching a particular topic. Here is an approach we can take to teaching light properties.

1. Instructor starts the first session on Light by darkening the room and lighting a candle. Then wonders out loud, what's going on, here in the room?

2. Students investigate the behavior of light from a source, for example they adjust the distance of a card from a light source and observe how the brightness of illumination changes. They explore other aspects similarly.

3. Students formulate and list all the properties of light they have discovered, and instructor and students together devise a pictorial 'ray' model of light to represent this behavior.

4. Students use this model to predict what areas of a screen will be lit or dark when an object blocks some of the light from a small source, and how the shadow size will change if the screen is moved further back for example.

5. Students reflect back on what the evidence was for each property of light.

Learning cycles – not a rigid sequence

Learning cycles are often shown as diagrams like those above. Although we show them as cycles with arrows, note that this does not imply a rigid step-by-step sequence! Nor are the phases 'compartments', separate and independent of each other – real science goes back and forth a lot, as does our own learning. There will also be direct links between any pair of phases. Note also that one may traverse the cycle more than once in developing a topic fully. Thus the phases and the cycle idea are not meant to imply some sort of rigid 'scientific method'. Rather, the phases and cycle representation are meant to bring out in a simple visual way the important aspects of doing and learning science, and how they all work together.

Note that some authors and educators may interpret the meanings of various phases differently, or may use variations on the names of phases. We adopt the original Karplus intent in our phase descriptions.

Advance organizers

An 'invitation to learn' phase at the start can include both an 'engage' activity and an 'advance organizer'. Engage can take many forms, but the idea is to awaken interest and curiosity, commonly through a memorable *specific instance.* This can be followed by an invitation of a quite different nature, namely an *'advance organizer'.* This provides a more general context and purpose for the topic to follow. That is, it raises the questions of interest and outlines what we want to know, why and how we might get there. It provides a big picture framework, of the forest if you like, without which students may just see one tree after another without knowing where they are going or why. The advance organizer lets them into the picture and the purpose, to better make sense of the details and specific activities to follow. It gives a sense of what it is all about.

When undertaking a journey of discovery, it helps to at least know why we are setting out and in what direction!

The concept of advance organizer is due to David Ausubel and is often misunderstood. Note that an advance organizer is *not* meant to be a 'summary' or outline of what is to come. (Especially not in an inquiry-based course!) Instead, it acts to frame and give purpose to what is to come. In that sense it prepares the way and the learner.

The absence of any sort of organizer, even implicit, is one of the major shortcomings of much instruction. Students often only realize at the end what it was all about, finally making sense of it if they are lucky. Otherwise they will just try to memorize the many details that were rushed by them.

Even a short introductory paragraph, or sentences inserted along the way during topic development, can serve this organizing function.

Instructional approaches

The learning cycle approach to science instruction aims to incorporate the various facets of science and of instruction in a balanced way.

Note that many conventional courses generally omit the exploration and concept development phases, and start by 'providing' knowledge directly. That is, they deliver 'ready-made-science' to the students. Some might argue that this is the simplest and most efficient kind of instruction, with the instructor 'transmitting the facts' for the students to 'receive', as well as explaining them. However, there are problems with this 'knowledge transmission' model. Firstly, it comprises just the ready-made-science face; this is a misrepresentation of real science and has an undesirable effect on the perceptions of students and instructors alike about the nature of science. Secondly, we know that learners have to construct their own understanding and make meaning of things for themselves, rather than being able to simply ''absorb' knowledge intact.

In our approach, all three faces of science are emphasized. To achieve this in instruction we adopt the 5-E learning cycle model for developing most topics. The cycle also incorporates some of what we know about learning.

Note also that during instruction, students may not always be aware of the status and purpose of a particular activity they are busy with, unless there is a clear structure in which to locate it. We aim to provide that, for both students and instructors, by making the cycle phases explicit. Thus everyone should be aware whether an activity is about exploring, seeking relationships, formulating principles, testing them, or applying them, etc. and why.

Building the processes of science structurally into the curriculum as an explicit framework for developing subject matter is a key factor influencing both instructor practice and student understanding of the nature of science. Our course is explicitly inquiry-based, going from process to product to power. This scientific inquiry framework should also help learners become familiar with the discourse of the scientific community by engaging in 'doing science', albeit in a modest way.

The investigative theme also acts as a course 'storyline', involving observing, posing questions of nature, investigating, inventing concepts and theories, and then applying them in new situations. This kind of purposeful narrative can make a course interesting, more so than if it is just a compilation of factual knowledge.

Hammer and Elby note that many students appear to view scientific knowledge as coming from authority, but that nevertheless even small children have the ability to understand knowledge as *invented*. This is the aspect we wish to promote, despite the contrary effects of much conventional science instruction.

Note that foregrounding 'development of knowledge' does not mean downplaying the 'application of knowledge' component in any way. Application of knowledge traditionally involves problem-solving to a large extent, and active engagement with problems is an important part of science learning.

1-5 "POWERFUL IDEAS"

'Powerful ideas' in science – a theme of this course

A goal and characteristic of science is to explain a variety of situations in terms of relatively few fundamental concepts and principles. Accordingly our inquiry-based physics course uses an ongoing 'powerful ideas' theme throughout. In each area of science we focus on the big ideas which can account for a range of phenomena. They are powerful for explaining, predicting, and solving problems.

We hope that you will come to appreciate this perspective, in both learning physics fundamentals and applying them to solve problems.

This focus on principles is also a more productive way to address 'misconceptions' than is possible with approaches emphasizing 'facts'. Thinking out situations in terms of underlying models and powerful ideas is preferable to recall (or mis-recall) of taught knowledge.

Note that some powerful ideas are also 'big ideas' in that they are overarching or unifying ideas for a whole field, while others may be smaller or more targeted in that they apply to a specific more limited area. In either case however, the ideas are 'powerful' in our hands, for understanding the world and solving problems.

Powerful ideas may be of various kinds, e.g. theories, principles, laws, models or concepts. They are what the human intellect has constructed over thousands of years of investigating and trying to understand our universe.

Examples of powerful ideas in science

We will develop many powerful ideas as we go through the physics course, but here we list a selection of powerful ideas you may have come across in science, to give you an idea of what we mean.

Physics
Big ideas: • Newton's laws of motion • Conservation laws (conservation of energy, conservation of momentum etc). • The ray model of light. •
Add your own . . . . . .
Specific area ideas: The image distance rule for a plane mirror. • Add your own . . . . .

Biology
Big ideas: The theory of evolution. • Add your own . . . . .
Specific area ideas: • Add your own . . . . .

Chemistry
Big ideas: The atomic nature of matter. • Add your own . . . .
Specific area ideas: • Add your own . . . . .

Earth Sciences
Big ideas: • Add your own . . . .
Specific area ideas: • Add your own . . . . .

Various classes of powerful ideas

Not all powerful ideas are created equal! Some are fundamental, applying universally as far as we know. Others are secondary in that they can be derived from the fundamental ideas. Others are 'special case' powerful ideas, which apply only to specified cases or under special conditions. These are obviously less powerful, but are very useful in their domain. Further types of ideas are included in the list below. The instructor will discuss them briefly and they will become more meaningful when we encounter examples during the course.

Various classes of powerful ideas

o Primary Principles (fundamental)	PP
o Secondary Principles (derived)	SP
o Special-Case Principles	SCP
o Powerful Concepts	PC
o Powerful Models	PM
o Powerful Mathematical Ideas	PMI

Identifying, naming and listing these serves to reflect our course emphasis on theories, principles, concepts and models rather than simply on 'facts'.

Notes

18

1-6 LEARNING AND PROBLEM SOLVING

Understanding science and solving problems involve *human cognition.* Expertise is not just about the science, but also about how we *think.* In the next sections we talk about aspects of cognition important in physics, and their implications for teaching, learning and problem-solving.

MODES OF REASONING IN PHYSICS

INTRODUCTION

What kinds of thinking are involved in learning and understanding physics? What reasoning modes and knowledge bases are important in solving physics problems?

You might think that a scientific approach to solving problems would consist of systematic application of appropriate physics principles to the problem at hand. And in some sense you would be correct – but only partly so. The thought processes involved in tackling problems are actually much richer and more complex than this.

And what about the *teaching* of problem-solving? How is this best done? Teaching physics problem-solving, (if done explicitly at all) often consists of 'going over' worked-out final solutions as a stepwise application of physics principles. While this does represent the physics structure of the final solution, it does not reflect how people actually think when tackling problems. We find that both learners and experts use multiple modes of cognition in problem-solving. Three significant modes are: *principle-based reasoning, case-based reasoning* and *experiential-intuitive reasoning.* We discuss each of them below.

1. PRINCIPLE-BASED REASONING (PBR)

> In *principle-based reasoning*, problem solutions are obtained by a systematic application of physics concepts and principles.

Discussion

A goal and characteristic of science is to explain a variety of situations in terms of relatively few fundamental concepts and principles. Accordingly our inquiry-based physics course uses an ongoing 'powerful ideas' theme throughout. We hope that students will come to appreciate this perspective in both learning physics principles and applying them to solve problems.

It is clear that many entering students are not used to thinking and working via fundamentals. However, we hope that our emphasis on 'powerful ideas' does have an effect during the course, and that students learn to work in principled fashion applying those ideas.

The instructor can provide examples, now and during the course.

Books and instructors usually teach this 'principled' mode of reasoning by working carefully through 'model' solutions, discussing each step.

While this does represent the physics structure of the final solution, and the method is indeed powerful, it turns out that it does not reflect how people actually think when tackling problems! Or rather, it only tells half the story. For a more complete and realistic story, we now look at two other significant modes of reasoning.

2. CASE-BASED REASONING (CBR)

In case-based reasoning, we try to understand new cases by association with similar cases that we have already encountered. In problem solving, we recall how we solved related problems previously, or may even recall the solution. In this mode of reasoning we thus draw on our compiled case knowledge, rather than starting from scratch for each new situation. However, we must make to take any differences between the situations into account. We should also crosscheck against principle-based reasoning.

Discussion

We noted above that working from principles is powerful in solving science problems (which is why those principles are called 'powerful ideas'). However, this does not mean that we have to approach every problem from the very beginning in this way.

When confronted with a new case, it is very effective to recall the essential features of similar cases, i.e. to build on the previous knowledge you have developed about this area, as a basis for starting to think about the new case. That is, you can build to some extent on knowledge previously constructed – as long you take into account both similarities and differences between cases.

This mode of reasoning, based on knowledge of relevant cases, is called *"case-based reasoning'* or CBR. It involves thinking by association and analogy, and recalling previously compiled knowledge and cases.

The knowledge accessed can be of different types, e.g. details of a particular case dealt with previously, or a package of underlying knowledge and procedures, i.e. a useful 'pre-fabricated part'. The latter might for instance be a theoretical part based on principles and procedures, and retrieved as a whole chunk.

An example might be a diagram of two similar triangles for obtaining side ratios, useful in various shadow problems. Another would be a vector subtraction diagram used in deriving acceleration in circular motion, or in finding the momentum change in a collision. The instructor can provide other examples, now and during the course.

Of course any conclusions suggested by case-based reasoning should be checked against principle-based reasoning - ideally we should use CBR and PBR in conjunction with each other.

In a very simplistic form of case-based reasoning, students may try "mapping across" results from a remembered case directly onto a new case, instead of proceeding from principles. They merely try to reproduce an answer they recall, instead of trying to *understand* the new situation and think scientifically. Thus case-based reasoning can be done well or badly, and if done badly can lead you astray.

Thinking by association seems to be a natural human tendency. It has pros and cons, in both everyday and scientific contexts. Working by *association* with known cases, instead of via basic principles, though it has its risks, is not broadly undesirable. It works well in everyday situations and is quick. Used correctly it can be very productive scientifically too. Tapping into previously compiled case knowledge is very efficient, compared to working everything out anew from scratch every time a new situation is encountered. Experts do this as part of their repertoire! They have a rich store of compiled knowledge and particular cases to draw upon. In fact this is one of the features of expertise, and indeed of structural understanding of a topic, and it is hard to imagine operating without it – everything would be cumbersome. But when drawing on case knowledge, experts are aware of what they are doing, know the status, applicability conditions and limitations of that knowledge, and are able to crosscheck against basics. They also recognize that various cases are just particular instances arising from the same underlying principles or mechanisms.

For problems in a particular area, note that once one has solved one or two cases it is hard to imagine solving a variant *without* drawing on the mental picture already in place! This picture represents not just the result but also the method, and encapsulates one's previous construction of understanding of such situations. However this kind of compiled case knowledge, where the essence is understood as well, is different from the simplistic 'mapping across' of results from one case to another, which some students try, if they do not perceive both the similarities and differences between the cases.

3. EXPERIENTIAL-INTUITIVE REASONING (EIR)

In situations related to everyday life, we all have intuitive ideas about the world and how things behave, based on our own experiences. This is the basis of what we can call 'experiential-intuitive reasoning'. However this is not so much a conscious or deliberate mode of reasoning as it is a spontaneous intuition. It works reasonably well in situations close to experience but may lead to error if applied inappropriately to other situations. Experiential-intuitive reasoning may seem at odds to scientific ideas if differences in situations or concept meanings are not recognized, giving rise to so-called 'misconceptions', while in other cases it may assist conceptual understanding.

Discussion

Our intuitive responses can be interpreted as arising from so-called 'phenomenological primitives' (or p-prims for short), i.e. notions which are 'primitive', arising directly out of experience, or perhaps even instinctive, rather than being deliberately thought out. Different p-prims seem to be cued by different situations.

Two examples of p-prims involving motion are: i. the primitive experiential notion that force is required to keep an object (like a box) moving along the floor, ii. the primitive experiential notion that objects (like balls or bicycles) just naturally keep going unless there is something opposing them to stop them. You will notice that these seem to be contradictory p-prims! Yet they both capture some aspect of the real experienced world. The situations are different, and we need science to understand the differences.

Experiential-intuitive reasoning based on such primitives seems more prevalent in mechanics than in optics, perhaps because we all have direct tactile experience with the mechanical world. However we will encounter several examples of optics p-prims.

P-prims affect peoples' learning of science, depending on whether scientific and everyday ideas seem in accordance or at odds. In some situations, appropriate p-prims can *aid* conceptual understanding of science, in which case we call them 'resources' for leaning. In other situations, inappropriate p-prims may be cued, and hinder acquisition of the correct scientific ideas, leading to so-called 'misconceptions'. We prefer not to use this term, but rather to speak of 'conceptions' which may be used appropriately or not, and so can either help or hinder understanding

DISCUSSION OF REASONING AND IMPLICATIONS FOR INSTRUCTION

We find that learners and experts alike invoke multiple modes of cognition in physics problem-solving. Case-based reasoning, drawing extensively on pre-compiled knowledge, is pervasive during the process, though this may not be evident in final constructed solutions and explanations.

Significantly, working linearly and systematically from basic principles is not usually the primary mode of initial thinking of either learners or experts, although it plays a crucial role in working out the final solution. It is the way that we conventionally present model solutions, 'after the event', in teaching and textbooks.

The issue of **transfer** in learning is also relevant here. That is, the transfer of something learned in one situation across to another situation. Of course if transfer is to occur, then ideally students should carry across deep features (underlying principles and procedures), rather than surface features or particular results.

Students need to be aware of how science works, and also that scientific thinking is different from everyday thinking in certain ways. They need to appreciate the status of both general principles and particular examples, and know that applicability and generalizability need to be kept in mind. Such *epistemological* knowledge, though often tacit, is as important as declarative and procedural knowledge of the subject matter.

Implications of all this for more effective science education are that instruction should reflect what we know about real cognition and the nature of expertise. This includes teaching case-based as well as principle-based reasoning and making such thinking 'visible.' Simply re-teaching the physics without the cognition will not work effectively.

Thus in this course we tackle these cognitive issues in a number of ways: identifying 'sub-assemblies' of knowledge for each topic: presenting two or more case variations to aid discrimination of commonalties and differences; extensive use of problems in teaching, as formative assessment for learning; promoting metacognition and explicit reflection; and modeling these aspects in teaching and in the design of learning tasks.

TOPIC KNOWLEDGE SCHEMATA

Understanding a physics topic involves several related types of knowledge, e.g. knowledge of the phenomenon, its behavior, the basic concepts, principles and procedures involved, features, typical problems and various particular cases. After mastering a topic, we can store some of this acquired knowledge in memory in partly *assembled* form, i.e. in what we might call 'knowledge subassemblies'. Then the next time we encounter the topic or solve a problem, we can quickly retrieve an appropriate subassembly, rather than having to think entirely from scratch. We will use the term *schema* for such a topic knowledge subassembly. (Plural *schemata*). People use schemata, often unconsciously, to organize current knowledge and provide a framework for future understanding.

An example of a schema might be a visualized ray diagram representing the basics of image formation for refraction at an interface ('apparent depth').

Since we think by *association* to a fair extent, it might as well be by association with the best correct knowledge for the purpose! Hence we will *deliberately* devise good schemata for topics in this course. Such schemata can become part of our mental toolbox, to draw on as needed in understanding a topic or solving problems.

There are various types of schemata that can represent important aspects of a topic and serve as tools for thinking. When we reflect back on a topic in the 'evaluation' phase of a learning cycle, we can identify the essence and suggest schemata to represent it. Diagrams are useful ideal for this and are readily remembered.

Various possible types of schemata are listed below.

Useful types of topic schemata/subassemblies

➤ Phenomenon diagram

Ask yourself: what is the *essence* of the basic phenomenon? (E.g. of reflection or refraction). Next, can you devise a diagram to represent this? Ideally it should be abstract enough to represent the basic features without unwarranted detail. Points to note can be written alongside. An example would be a diagram of a light ray being reflected by a mirror, or a ray being refracted as it goes from one medium to another.

➤ Behavior (variation) diagram

It is also useful to have knowledge of the *behavior* of the system. That is, how things *vary* if you change something, or how one thing *depends* on another. This can often be shown on a diagram, accompanied by notable points. An example would be a comparative diagram showing rays incident at different angles being refracted.

➤ Principles and processes diagram

Here we try to represent the underlying science of the situation. Thus this will a more theoretical type of diagram, showing the principles and processes involved. An example would be a ray diagram showing how an image is formed by reflection of diverging light rays by a mirror.

> ➢ Features diagram

A feature diagram shows just the main features of a case, not necessarily the principles. An example would be the feature that the image is the same distance behind a plane mirror as the source is in front. The feature diagram just shows the result, and does not try to represent *why* this should be the case.

> ➢ Case comparison diagram

Where distinct cases can arise it is useful to represent them comparatively. An example would be a comparative set of diagrams showing distinct cases of image formation by a concave mirror.

Notes

Subassembly. A knowledge schema can also usefully be called a of knowledge *subassembly*. The *sub* indicates that it is a *component* or *part* of the whole knowledge needed, and *assembly* indicates that it has been pre-assembled (in your mind) from previous experience.

Prefabricated house analogy An analogy would be a partially pre-fabricated house – the house is put together from sub-assemblies already available, e.g. windows, panels, doors, even roof structures, etc. Thus a building is rarely constructed completely from scratch, i.e. from bricks, lumber, glass, etc, (although of course it can be). On the whole it is efficient to use pre-fabricated parts for 'standard' items.

Cognitive schemata or subassemblies. It turns out that the same is true of cognition. First, we build knowledge schemata from cases we encounter while learning. Then, when tackling a new problem, we do not always have to work out the solution from scratch from basic principles – though of course it can be done that way. We can draw on our previously compiled subassemblies of case knowledge. Experts do it, novices do it, we all do it! It is efficient to do this. The subassemblies already incorporate the essence of what we have learned before. But of course we should always check our pre-assembled bits of knowledge against basic principles, and also make sure the knowledge parts are appropriate to the new situation – just as pre-fabricated windows must be the right ones for the new house!

Abstraction of essence. Ideally a topic knowledge subassembly will contain the essence of our understanding of a case i.e. the situation, the principles, the procedures, and the characteristic features. It will represent all this in a fairly abstracted way, so that it will be relevant to this *type* of situation whatever the surface details of particular cases.

Mental images. You can form mental images of these diagrams, to access whenever you think of the topic. They contain within them the knowledge and understanding you have already built up in studying the topic. Note that when in solving a problem you find yourself recalling a diagram in your textbook or notes, you are in effect retrieving a topic subassembly! One into which you have previously put a lot of thought, so you can now benefit from that pre-compiled thought.

All this will make more concrete sense once we are further in the course and have examples to work with! However we outline it here in advance, so that you are aware of the idea, and because the concept of schema or subassembly will be applicable to all topics.

1-7 COMPARISON WITH CONVENTIONAL TEXTBOOKS

This book takes particular approaches to a number of science education issues, which may not be the conventional. For example it takes a *scientific inquiry* approach to developing physics topics, rather than presenting known physics content. This is embedded structurally using a learning cycle approach. The pedagogy is one of *guided inquiry*, rather than 'direct' presentation. An ongoing epistemic theme is that of 'powerful ideas', emphasizing underlying fundamental principles, in both learning and problem solving. The application side is strong, with sets of specially designed problems as 'assessment for learning'. Issues of cognition and epistemology are integrated with the physics content and problems, as are learning and teaching issues.

Sometimes, in order to appreciate what an approach **is**, it helps to compare with what it is **not**. Therefore, after working through any chapter in this book, you might like compare the treatment with a few conventional physics textbooks. The instructor could also provide comparative examples of inquiry and direct approaches for various topics. For each approach, you can identify its characteristics, the picture of 'science' that it conveys, and the instructional approach.

You might like to reflect on whether the conventional presentation of science by providing the known 'conclusions' of science at the start is akin to telling the reader "who dunnit" right at the beginning of a mystery novel. For our part, we prefer to offer science as a fascinating mystery story. The 'solution' to the mystery comes after 'detectives' have asked questions, investigated, gathered clues, tested hypotheses and finally solved the mystery – based on evidence, thinking, and insights!

Although this book is designed in terms of inquiry and learning cycles, conventional curricula and textbooks are not. How then can instructors best teach from available conventional materials, or students learn from them? With awareness of the issue, instructors can put in or generate the missing aspects to some extent. Thus if available materials have a 'direct' approach, a teacher can enhance this by adding an inquiry front end and a reflection back end. If the application aspect is weak, the teacher can enhance existing exercises or devise new problems. On the other hand, it may be more satisfactory, though time-consuming, to seek or write materials geared to the desired approach from the start.

Students too, faced with direct materials or instruction, can consciously re-cast this to some extent by framing things themselves from a scientific inquiry perspective, asking the appropriate inquiry questions and trying to answer them. Problems too can be enriched with conceptual aspects. Of course all this enhancement for deep understanding is an additional study burden, but it is nevertheless an extremely valuable talent to develop, since much of the material we encounter in life will be formulated in 'knowledge presentation' fashion.

Chapter 2

LIGHT: EXPLORING ITS BEHAVIOR AND DEVELOPING A MODEL

SECTIONS

PROBLEMS

2-1 PRELUDE: EVERYDAY EXPERIENCE OF LIGHT AND QUESTIONS ARISING

Light!

This course is about light and its properties.

Let's start with a simple experience: darken the room completely – this is the *absence* of light!

Then after a minute, light a single candle. See what happens in the room. You are now experiencing the phenomenon of interest – light!

It may seem commonplace, so you may not have thought about it much. However as scientists we will want to study light systematically, investigate its behavior, discover its properties and devise a model to represent them.

A candle flame wavers and is not as tiny as we would like as a source to start our investigations. A simple alternative is a tiny bare 'maglite' flashlight bulb. Darken the room again and remove the reflector from a maglite, turning it on. This is the source we will use for our initial experiments.

LIGHT!

"It is better to light a candle than to curse the darkness"

Curiosity and questions

Observing light from this source should spark curiosity. What questions arise? What would you like to know about light? How could we go about investigating light to find out?

Reflect, discuss, and write down things you might like to find out. Groups can use a whiteboard.

Questions that arise

Your own experiences of light

We all have *experience* of light from everyday life. We thus have some knowledge about its behavior, even though we may not have thought about it much. So, before reading further, let us start from our existing experiences, as a prelude to studying light further. What you know about light from your *own* experiences with it? (Not what you may have been *told* about it). Reflect, discuss, and list alongside.

From this, we get an idea of what we already know about light from common experience –

What you know about light from your own experiences

and perhaps an idea of what else we might *want* to know! More questions will arise as we begin exploring light further.

Taught facts about light or our own investigations?

Most people have been taught something about light at school, or read about it. However we prefer not to review such 'taught facts" about light. Instead, we will investigate light behavior ourselves and reach our own conclusions from the evidence. Thus in the next section we start our first simple but systematic explorations.

2-2 EXPLORING THE BEHAVIOR OF LIGHT

Light from a point source: exploring behavior, formulating properties and developing a model

Our approach to studying light

Think of this as starting a scientific journey to find out about light. The way we approach our study of light is to start simple – with just a small 'point' source of light. Surprising as it may seem, just using a single tiny source we can find out quite a lot about light behavior! We can explore how the light behaves in various situations, and list the properties we discover. In this way we aim to develop a *model* for light behavior, which embodies these properties. Then we will test some implications of our model, and finally use the model to solve problems in new situations.

Later we will turn to more complex sources and situations, but we will start as simple as we can. Physics usually starts with the simplest situations, working toward the more complex.

Posing questions and starting to explore

Let us start to investigate properties of light from a point source.

Set up a small light source (e.g. the tiny bulb in a 'maglite' with the reflector removed). Turn it on.

Focus questions

We first pose questions about the behavior of light from this tiny point source. Here are some possible basic questions:

Point source of light

- A. Does light 'travel'?
- B. Does it travel in straight lines?
- C. If light travels, how fast?
- D. What about directions of emission?
- E. Does light intensity vary with distance?
- F. Can light travel through nothing (vacuum)? Through a solid or liquid?

We will explore these one by one. From our findings, we will aim to develop a model for light, that incorporates the various properties that we discover.

A. DOES LIGHT TRAVEL? OR IS IT SIMPLY 'PRESENT' IN SPACE? - AND HOW CAN WE CHECK THIS?

Focus question

Does light *travel*? Or is it simply 'present' in space when a light source is on? We need to do some exploring and thinking to find out.

Trying to observe or measure if light travels

Start with a simple trial: switch on an electric light bulb. Does light seem to be everywhere in the room immediately? Or does it travel out from the source, taking time to do so? Try it a few times and observe. Think and discuss.

Can we reach any conclusion yet, from this simple first trial? Can your trial distinguish between light being present instantaneously and light traveling extremely fast?

First observation attempt

A different approach to answering the question

It is difficult in class to find out whether light travels or not by trying to measure travel time. Is there some clever alternative way we could get evidence for traveling or not? Without having to observe the traveling or to measure speed at all. Think, discuss and suggest. Check your ideas with an instructor.

Then try out your test.

What do you observe and conclude?

Devise a way of testing
Try it and report

Our first knowledge claim: a powerful idea

Based on this evidence, put forward a first *knowledge claim* about light.

We have generated our first "powerful idea' about the behavior of light!

Note:
It is creative to have found out whether or not light travels, without having to actually observe it traveling!

First knowledge claim about light
Evidence

B. DOES LIGHT TRAVEL IN STRAIGHT LINES?

Focus question

If light travels, *how* does it travel? In straight lines? Or does it perhaps flow around round objects, like air or water would? And how can we find out?

Design an exploration and try it

How can we explore this question, in a simple way? Any ideas?

Think and discuss. What would *you* do? (There may be several possibilities).

Then try it out, observe, and state what you find and conclude.

Devise a way of testing
Try it and report

Hence state another knowledge claim

State your findings as a knowledge claim about light, and state the evidence you have for it.

Second knowledge claim about light
Evidence

Our second powerful idea

This is our second 'Powerful Idea' about light.

Combined statement

Note that people often combine the first two powerful ideas into one statement about light.

Statement:

Although this is concise, note that there are really two properties involved and we may prefer to think of them separately.

Additional demonstration by instructor

At this stage yu can try additional demonstrations to illustrate this property of light, by making light beams 'visible' using dust or spray mist.

Is this property of light 'obvious' or not?

In case you think it seems 'obvious' that light must travel in straight lines, and thus there is no need to check, consider a similar question about *sound* – does sound travel in straight lines? And how can we check this out? Then try it! Report.

So, it's not clear that everything must travel in straight lines! Science needs to investigate, and base any conclusions on evidence.

Light properties in other circumstances – wave-like behavior!

From our investigations so far, we can conclude that light travels in straight lines – in the situations we explored. We cannot go beyond that and conclude that it does so in all situations, without further investigation. We will not do that here, but it is important to be aware that in certain situations we find that light has another aspect to its behavior – it exhibits wave-like properties too! For example it can actually 'bend' slightly around obstacles. This effect us usually only noticeable on a very small scale. That that is the subject of other investigations, and does not affect the large-scale behavior that we are studying here. Thus as far as this course on geometrical optics is concerned, we assume light travels in straight lines.

Reflection and discussion

Reflect on this section, the approach taken and your own learning and thinking, and write down insights and useful ideas.

C. IF LIGHT TRAVELS, HOW FAST?

Focus questions

If light travels, how fast does it travel?
Or is it instantaneous perhaps?
And how could we find out?

Designing an experiment

Suppose we want to find out: i. whether light travels instantaneously or not, and ii. if not, what is its speed. How could we go about this task? Think a moment.

As a preliminary, ask yourself how you would measure the speed of some *ordinary* moving object, such as a bicycle traveling along the street or a toy car moving on the floor. How would you set things up, what would you measure, and how would you calculate speed? Write it up.

Now, how might we design an experiment to measure the speed of *light*, at least in principle? Think, discuss and suggest a design.

Trying it

Then try to measure light speed (as best you can) in the lab. Have fun trying, even if you run into problems and don't succeed! We only have limited equipment, such as stopwatches and tape measures, so if light travels very fast it will not be surprising if we can't manage to measure its speed.

> Suggest, in principle, a way of testing whether light is instantaneous or not.

> Suggest, in principle, a way of measuring the speed of travel of light
>
> Try it and report :)

Initial finding

From your rough attempt above with simple equipment, what is the most that you can conclude so far about the speed of light?

Statement:

Though limited, this is some knowledge after all!

Discussion, reflection etc

HISTORICAL ATTEMPTS TO MEASURE THE SPEED OF LIGHT

We can look back historically to see how scientists over the years tried to measure the speed of light. We start with Galileo.

Galileo's method for measuring the speed of light

From ancient times people have been intrigued about whether light was instantaneous (infinite speed) or whether it traveled with some fast but finite speed. If the latter, it certainly travels fast enough to make it very difficult to measure.

Galileo Galilei was one of the first to suggest a method of measuring the speed of light. Remember that in his day they had only crude timing devices, certainly not stopwatches or anything more precise. Galileo actually used his own pulse as a timing device for some of his motion experiments!

Task: Find Galileo's method and writings

As an assignment, search the library or web or for *Galileo's method* of attempting to measure the speed of light. There will be various accounts of it including Galileo's own writings. Print and save an account that you like, but also find Galileo's *own* account in his own words, from his book 'Dialogue Concerning Two New Sciences' – English translation. (Ask the instructor if you have difficulty finding it). Note that Galileo writes his account as a narrative dialogue between three people, Salviati, Sagredo and Simplicio.

Print what you find and bring to class. Groups can present what they have found. We will discuss the merits and drawbacks of Galileo's method.

Task: Chronology of measurements of the speed of light

In searching for Galileo's method you will likely find a chronological account of various methods of finding the speed of light, from ancient up to modern times, getting more and more accurate over the years as science and technology advanced. You might like to compile these documents to accompany this section, for interest's sake, and bring to class to share.

Which travels faster, light or sound?   A comparison test

Focus question

We now know that both light and sound travel, so an interesting question arises – *which travels faster?*

Think of a method of comparing

With our limited equipment it is virtually impossible to *measure* the speed of light. However if we only wish to *compare* the speeds of light and sound, i.e. find out which is faster, we can do it quite simply without needing to measure the speed of either!

The idea is simple: we need to make something happen that we can both see and hear from a distance. Think and discuss: what situation could you set up, what would you do, and how you could tell whether light or sound was faster?

Suggest a method to find whether light or sound travels faster

Class experiment outdoors

The experiment is done as an outdoor class activity. Ideas for the design of the experiment will be discussed, and the instructor will provide simple equipment and guidance.

Then do the experiment.

'Fact picture' of the activity:

'Abstract diagram' to represent the essence:

Conclusion

State your findings and conclusion.

Extension task:
To get a rough estimate of the speed of sound

It turns out that sound travels pretty fast by everyday standards but slow enough that you can do a rough measurement of its speed, if you use a travel distance of one or two hundred meters and estimate what fraction of a second the sound takes to reach you. Note that you can probably *estimate* a third or half of a second without using a stopwatch, once you get a sense of what a second is like. The instructor will discuss. The point of this task is to understand the method, and get a very rough *estimate* of sound speed, rather than trying for precision, which is not the point here.

Try it. Pace off the distance roughly, estimate the time taken for the sound to reach you, as a rough fraction of a second, and hence get an estimate of the speed of sound!

Working

Approximate distance traveled: . . . .

Estimated time taken: . . . .

Use verbal logic and mental arithmetic to estimate the speed of sound, like this:

Sound takes . . . . . of a second to travel . . . . . meters,

Therefore in **one** second it would travel . . . . meters.

So the speed of sound is roughly . . . . . . meters per second.

Discussion, reflection, notes, etc

• Note that this is very approximate, depending on how good you are at estimating a fraction of a second! But that's all we want for the moment, just a rough idea.

EXAMPLE PROBLEM – lightning and thunder

During a thunderstorm, you see a lightning flash, and immediately start counting off seconds as best you can. Suppose you hear the thunder about three seconds after you see the lightning, by this rough method. And you know the speed of sound. About how far away was the lightning strike? Do this by a series of logical statements, rather than a 'formula'. Should you be afraid of being struck by lightning on this occasion?

D. DIRECTIONS OF LIGHT EMITTED FROM A POINT SOURCE

We continue our study of the behavior of light emitted from a point source. This time, we ask about the directions of light emission.

Focus questions

What are the *directions* of light emission from a point source, e.g. our small maglite bulb? Is light emitted in all directions, or preferentially in some directions? And how can we find out?

Design a simple test

Design a very simple test, to find if light is emitted equally in all directions or not. Think, discuss and propose what to try.

Try it and observe

Suggested tests
Try it and report

Hence propose a knowledge claim

Based on your observations, what property of light emission can we infer? Formulate a knowledge claim, as another 'powerful idea' about light behavior.

Counter-example?
What about the directional beam of light from a flashlight?

Does a flashlight beam, concentrated in one direction, contradict our claim about directions of light emission? Discuss and explain.

By the way, this is our first example of *manipulating* or *controlling* light, in this case by means of a reflector. Look at the reflector around the flashlight bulb, and how it is shaped.

E. DOES LIGHT INTENSITY DEPEND ON DISTANCE?

We continue our study of light, by looking at another aspect of light behavior. At the moment this will be qualitative only.

Focus question

Does the intensity of light illumination vary with *distance* away from a point source? And how can we investigate to find out?

Designing a test

How could you explore this in a simple way? Design a simple test, then try it.

Exploring

Then try it out, and report findings.

> Suggested simple qualitative test
>
>
> Try it and report :)

Formulation of knowledge claim

Formulate your findings as another knowledge claim about light – a further powerful idea!

Any ideas about **why** intensity changes as distance changes?

This will be the subject of another chapter, but for the moment let's simply brainstorm ideas about *why* light intensity decreases as we get further away.

Think of as many possible ideas as you can, gather more ideas from others, and list them all on the board!

We will not be investigating yet to find out which ideas work or not, but this is a good place to start wondering.

> Suggest possible explanations for the change

Note

This activity has been qualitative so far. We have discovered the general type of behavior, but have not (yet) made numerical measurements to find out quantitatively how intensity varies with distance.

F. THE MEDIUM THROUGH WHICH LIGHT TRAVELS

1. Can light travel through *nothing* (a vacuum)?
2. Can it travel through a solid or a liquid?

Question 1

Can light travel through nothing (a vacuum)? Or does it need a 'medium' (some substance) to travel in?

Design a test

How could we find out, either by an experiment and/or by observations in nature? Think and suggest.

Suggested tests or observations:

Experiment

If we want to investigate this experimentally in the lab, we need to make a vacuum in an enclosed space – or at least get rid of as much air as possible. The instructor will show you a way to do this, using some quite simple apparatus. Try it out to see how it works.

Then shine a light beam through the air in the bulb and predict – as air is gradually pumped out of the space, what will happen to the light? Will it get dimmer until it no longer passes through, or will it pass through just as before?

Then try it – does light pass through the evacuated space?

Apparatus setup

State your conclusion, as another discovered property of light:

Aside: the case of sound

In case someone might think this property of light might be expected or 'taken for granted', ask the same question about *sound*: can *sound* travel through nothing (vacuum)?

What do you think?

And how would you investigate to find out?

Although this is not a unit on sound, maybe the instructor will be able to demonstrate this or show a video, just for interest.

Sound setup

Question 2

Can light travel through a *solid*?
And through a *liquid*?

Design a test

How could we find out, either by an experiment and/or by observations in nature? Think and suggest.

Experiment

Try some simple experiments and/or observations to test this.

Decribe your experiments, then report.

Suggested tests or observations:

Test it and report back:

Notes

1. Although we are familiar with the property of light that it can 'pass through' certain materials, this is puzzling when you think about it. Light can pass through certain substances like glass, which are very hard and solid to the touch (certainly your hand won't go through) with no trouble at all. Amazing! What IS light anyway, that it can do this? To understand this we would have to find out a lot more about light, and also about the structure of the material it is passing through. Science has made good progress on both of these fronts! But at this stage we are in no position to tackle them, we are just exploring some of light's basic properties, without aiming to explain them all.

2. Here is another interesting question: when light is traveling *inside* a substance, is it different in any way? We can't say from our simple investigation so far, but your may be interested in what scientists have found: the *speed* of light is different in different substances! This property gives rise to the phenomenon of refraction, or bending of light, which we investigate in a later chapter.

2-3 TAKING STOCK SO FAR:
KNOWLEDGE CLAIMS ABOUT LIGHT, WITH EVIDENCE

From a series of observations in this section we have developed several knowledge claims about light. List them in the table below, together with the evidence for each claim. (The first item is put in for you as an example).

Property of Light (Knowledge Claim)	Evidence
A. Light travels!	A. Light can be 'blocked'.
B.	B. . . . .
C	C. . . . .
D	D. . . . .
E	E. . . . .
F.	F. . . . .

Note that we have developed this knowledge about light from our *own exploring and thinking,* i.e. we have been 'making' science, rather than simply being 'told' the properties of light.

If these are ideas about light behavior are indeed 'powerful' then we should be able *apply* them to new situations, not just to those particular situations from which they were developed. We will be using these powerful ideas throughout the course, in different contexts.

2-4 DEVISING A MODEL TO REPRESENT LIGHT BEHAVIOR

Devising a model

So far we have produced a list of properties of light, by investigation. We can now try to devise a *model* to represent this light behavior, a model that incorporates these properties, or as many as possible.

In our case this will be a 'pictorial' model, where we use straight lines to represent light travel from a source. Do you have any ideas about how to do this? Suggest how we can represent light being emitted from a point source, and how this pictorial model can represent the various properties, i.e. that light travels; it travels in straight lines; it is emitted in all directions; and the light intensity decreases as we get further away. Draw your proposed model in the box below: Then say how each property is represented pictorially.

Pictorial model for light from a point source	State how each property is represented

Comment

This is our initial 'ray' model of light behavior. It can embody most of the properties so far, in a simple way. It is thus quite an efficient representation! Notice that we do not try representing the speed of light in this model – models can't do everything!

Such a model is useful, in that we can *think* about light in terms of it, and it can help explain or predict light behavior in new situations.

We use the model in the prediction tasks that follow.

2-5 Testing our model of light behavior

Introduction

If our model is valid and useful, it should enable us to predict what will happen in new or more complex situations. This can serve as a test of the model: we use the model to make predictions, and then try things out in practice and see if the prediction is confirmed. If so, we have confidence in the model, and if not we need to revise or rethink.

Let us try this in some test cases.

Test no. 1. Using the model to predict the illumination on a screen placed **far** from a source

Sketch the situation, draw a ray diagram showing light rays coming to the screen, and hence predict what the illumination will be like. Does this make sense?

Check against nature: is this what is actually observed? Try it.

Test no. 2. Using the model to predict the illumination on a screen placed **near** to a source

This time, place the screen very close to the small source. Draw a ray diagram and hence predict what the illumination on the screen will be like. Does this make sense?

Check against nature: is this what is actually observed? Try it.

Test 3: Using the model to predict the effect of **two** point sources far from a screen

Let us apply our model to a more complex situation. If we have **two** point sources of light near each other as in the sketch, what will be seen on a screen far away?

Sketch of two sources and screen

Simple wonderings

First, *before* using the model, just suggest some possibilities about what would be seen on the screen. Say what other people might think intuitively, and why.

Possible 'intuitive' ideas about what might be seen

Using the model to predict

Now use our *model* to predict, based on properties of light from the sources. Illustrate on the diagram above, showing how light coming from the two sources behaves.

Prediction based on model

Now try it out in practice, to test the predictions

Set up the two sources and a screen some distance from them, and observe the effect.

What do you observe on the screen? Take note of two things: the illumination *pattern* (if any), and the illumination *brightness* compared to that for a single source.

Is this a surprise, or expected?

Observations:

Note about the use of models to predict or explain

In this test we used the model to *predict* what we would observe. Alternatively, instead of predicting from the model, we could have first *observed* the effect, and then used the model to *explain* it. Thus models can be used either to *predict* new observations or to *explain* existing observations.

Any application of the model serves also as a further test of the model – each time we use a model we hope to get further confirmation that it works. Of course this might not always be the case – in science we sometimes find that our proposed models fail! Then it is back to the drawing board.

Test 4: Using the model to predicting what happens if light encounters objects and apertures

Suppose we have a light source and a screen, and there are some objects in between.

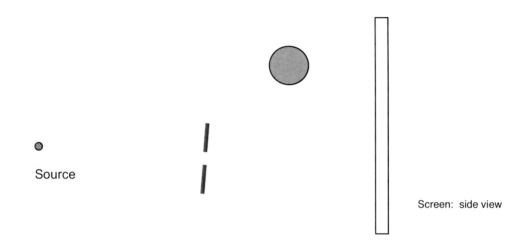

Source

Screen: side view

What will be seen on the screen? This is a new situation that our light model should be able to handle!   Try it and predict!

The idea was to use our pictorial model for a number of things:

- Use the model to represent how light is emitted from the source
- Using many rays, show how bright 'lit' regions arise on the screen (in side view).
- Using many rays, show why there will be 'dark' regions on the screen (i.e. shadows).

2-6 COGNITION IN THIS AREA
A 'knowledge schema' for light behavior

In Chapter 1 we introduced the concept of 'topic knowledge subassemblies' that could represent the essential features of a particular topic and serve as tools for thinking and problem solving. Various possibilities were phenomenon diagrams, behavior diagrams, principles and procedures diagrams, feature diagrams and case comparison diagrams. Please review that section of the chapter.

Now that we have studied *the behavior of light, listed properties and devise a model,* we look back on the essence of the topic, and ask – what would be useful subassemblies to encapsulate this knowledge?

It turns out that our ray model of light behavior already fulfils many of these functions, being already a pictorial diagram. Draw the model again alongside.

It can clearly serve as a *phenomenon* diagram and a *behavior* diagram. It could also serve as a *principles and procedures* diagram, if we think of *applying* it to various problems, e.g. to predict what will happen if there is object in the way of the light. Thus we already have a suitable knowledge subassembly for this topic.

You can form a *mental image* of this diagram, to recall whenever you need to think about light behavior and apply it to new situations or problems.

Ray model of light behavior/properties

Serves also as as a topic knowledge subassembly

2-7 PEOPLES' EVERYDAY IDEAS ABOUT LIGHT

Introduction and purpose

Both adults and children may have various ideas about light (and sound) and their speeds, or they may not have thought much about it at all. It is useful for a teacher to be aware of the various ideas (often tacit) that students may bring to the teaching/learning situation, along with the reasons that they may have for thinking that way.

Thus in this task, you interview people to find out what their thinking is about light and sound.

Investigation procedure and questions

How would you conduct this mini-investigation? Who would you ask, how would you introduce it, what would you demonstrate to people, what series of questions would you ask? Think, discuss and design the procedure.

The instructor will assist. Some possible interview situations and questions are given below to assist.

Possible demonstrations to show

- Switch on a lamp or flashlight, and have people observe.
- Hit something to make a sound from across the room.

Possible interview questions

- Does light travel or not? What do you think? Why? How do you know? Or don't you?
- Does it travel at some speed? If so, what sort of speed? Or is it instantaneous? How do you know?
- And how about sound? Does it travel? How? What speed? How do you know?
- Etc.
- Where do your ideas about all this come from, and how confident are you in them?

The investigation

Carry out the study with a few people, record data, analyze and interpret it, and write out your findings.

2-8 LIGHT EMITTED FROM E X T E N D E D SOURCES

Introduction:

What is an 'extended' source, as distinct from a point source?

We started our studies with the simplest possible case, i.e. light coming from a 'point' source. Physics deliberately starts simple!

Now, we turn to a more complex situation – light emitted from a larger ('extended') source. "Extended" means having some extent, and thus being of some shape also.

Line-shaped source. For example, an extended line-shaped source would look as shown alongside. The straight-filament light bulbs in the lab are good examples of line-shaped extended sources. A long fluorescent tube would be another example.

Extended 'line' source

Wondering

For the light emitted from a *line* source, what would be the effect on a screen nearby? How might the screen illumination look?

First, just suggest some possibilities, and say why you think these might be possibilities. Also state your own preferred prediction.

Suggest possible 'intuitive' ideas about how a screen would look if illuminated by a line source

Own idea:

Try it

Use the straight filament light bulb, place a screen at a fair distance from it, and note the effect. How does the screen look? Is this a surprise or not?

It seems like we are going to need a new or enhanced model, for light emission from extended sources, to explain this and other effects.

Toward a model for light emission from an extended source

Previously, for a point source, we devised a simple model for the emission of light. Now, we would like to have a suitable model for an *extended* source. Such a model would be useful it in understanding situations involving extended sources

Logically, we might suspect that an extended source is just like a whole series of tiny point sources. So we might view an extended source in terms of all the point sources, that make it up. Reasonable?

Hence i. Formulate a 'powerful idea' for light emission from an extended source, and ii. Devise and describe a pictorial model to represent light emission from an extended source.

Question: Can this model explain the screen illumination caused by a line source, the case that we investigated earlier?

> Powerful idea for extended sources:

> Pictorial model of light emission from an extended source:

Testing our extended-source model in a new situation
– predicting the effect of a V-shaped source

We developed our ideas about extended sources by considering a line-shaped source. To test our model further, let's apply it to make a prediction about a different shape of source. Light bulb filaments make convenient extended sources, and some common types have arc-shaped or V-shaped filament wires.

Get a light bulb with a V-shaped filament. Put a screen some distance away, but don't turn on the bulb yet. Before that, predict, *using the pictorial model,* what will be seen on the screen when you turn the V- filament on.

Then try it. Is your prediction confirmed or not?

Did the model help you make a scientifically based prediction?

If the extended source model was successful in predicting the result, this is further evidence for the validity of the model. If not, we would have to go back to the drawing board!

Note that many people, asked to predict the screen appearance, suggest that the screen illumination will be the same shape as the filament shape. Why do you think this is? After all, we all see the effect of filament light bulbs in everyday life.

> Sketch of setup
>
> Predict what will be seen on the screen
>
> Try it and report

PROBLEMS

The behavior of light emitted by a source

This application section might be called "idea power"! We have developed some fundamental scientific ideas about light behavior, and now we will *use* these ideas – to solve problems, to explain, and to predict.

In this chapter we systematically developed knowledge about light behavior, through an inquiry process. We formulated this knowledge as powerful ideas and powerful models. We did this for point sources and then expanded our ideas to extended sources.

But are you proficient yet in *using* this knowledge? This requires practice to develop expertise. So this section provides problems where you must *apply* your knowledge in a variety of situations – use it to explain, to predict and to solve problems. By doing this, not only do you learn to become proficient in problem-solving, but also consolidate and enhance your understanding of the basic physics. Note that if a person can't *use* their knowledge in new situations, they don't really understand fully! Their knowledge is said to be 'inert'.

Some objectives for this 'application of knowledge' section are as follows:

- To become competent in applying principles in both familiar and unfamiliar situations, in order to explain phenomena, solve problems and make predictions.

- To see how just a few basic ideas can be applied to *many* situations to solve a diverse range of problems. In that sense these are 'powerful ideas'.

The problems that follow are in categories, involving a single point source, two or more sources, and extended sources.

GENERAL

1. *Why start with a single tiny bulb?*

   Why did we start our investigation of light with just a tiny maglite bulb as source, rather than say an ordinary household lamp or a fluorescent tube? Would you do the same if you were to teach light in the classroom? Would you also explain to students *why* you do it that way?

LIGHT TRAVEL

2. *How do we know that…? What is the evidence for …?*

   It is stated that "light travels in straight lines". How do we know this? What evidence can you point to in everyday life, to support (or refute) this claim? How could we check it? If light did *not* travel in straight lines, what differences would there be?

   Note: a conventional textbook approach to this aspect of light would be simply to state it as a known fact, sending an implicit message that scientists have found it out and we should just learn their result, as already-made-science. The trouble is, when people do only that, they are generally unable to answer questions like How do we know that …? What is the evidence for ….?

3. *Does **sound** travel in straight lines?*

We have gathered some evidence that light travels in straight lines. In case you might think that this is rather simple and obvious, consider something else that travels – sound! Does *sound* travel in straight lines? Design an experiment to find out. Then try it at home and report.

Does this make you think again about whether it is 'obvious' that light should travel in straight lines? Does it point to the need for experimental evidence to back up claims?

4. *All directions*

Consider the statement: "Light comes out from a point source in all directions". How do we know this? How could we check it? What are a couple of pieces of evidence for this assertion?

5. *Flashlight source – an anomaly or not?*

Consider a flashlight – its light beam is very directional. Does this contradict our knowledge claim that "light travels in *all directions* from a point source"? Look at a flashlight and see how it is constructed, then explain.

SPEED OF LIGHT

6. *Galileo's proposed method for finding the speed of light*

Describe Galileo's lantern and mountaintop method of finding the speed of light. Say what measurements he hoped to take, and how he would calculate the speed of light from them.

Galileo's proposed method did not succeed. Why not? Here are some possible reasons: state whether each is true or false.

- His method was wrong in principle. T or F
- His method was fine in principle but impractical in practice. T or F
- Human reaction time is too long. T or F
- The time taken by light between mountaintops is much shorter than a person's reaction time. T or F
- The speed of light was too fast for this method. T or F
- If the speed of light had been much slower, e.g. like 100 mph, the method would have worked fine. T or F
- It would have worked if an adjustment had been made for human reaction time. T or F

7. *Time for light from the sun to reach earth*

The sun is 150,000,000 km from the earth. How long does sunlight take to reach us from the sun, if light travels 300,000 km in each second? Express your answer in minutes.

8. *Light from the nearest star – and the size of the universe*

The light from the nearest star takes 4 years to reach us! How far away is this star? Then find out its name.

Note: Travel times of years may at first seem amazing, considering how fast light travels! On the other hand, it just points up the scale of distances in our galaxy. In fact this is a very short distance in the universe! Distances to other stars and other galaxies are far greater than this, and light may take millions of years to reach us.

9. *Which travels faster, light or sound?*

Design an experiment to find out, for your class to try next lab period. Note it can be a 'comparison experiment', without necessarily measuring the speed of either.

10. *Workin' on the railroad – 170 meters away*

Railroad workers are pounding on steel rails 170 meters away from you. You estimate that hear the sound about half a second after you see the hammer hit the rail. From this information, estimate the speed that sound travels to you. What assumption do you have to make about the speed of light, compared to sound, to be able to work this problem?

11. *Workin' on the railroad – '2 seconds away'.*

Railroad workers are pounding on steel rails in the distance. You estimate that hear the sound about two seconds after you see the hammer hit the rail. You know that the speed of sound in air is about 330 meters each second. How far away are the railroad workers? What assumption do you have to make about the speed of light, compared to sound, to be able to work this problem?

12. *Thunder and lightning*

Homework and home experiment: During a thunderstorm you see a lightning flash in the distance and then count off 2 seconds before you hear the thunder. The speed of sound is 340 meters each second. How far away was the lightning strike? Do you need to be afraid?

13. *Light and sound experiment*

In our 'field trip' one student went some distance away and hit two pieces of wood together, and we heard the sound *after* seeing the boards hit. This experiment shows that:
    A.   Sound takes time to travel, but light is instantaneous.
    B.   Sound travels slower than light.

14. *Foghorn at a distance*

It is known that sound travels about 340 meters in every second. A Lake Michigan ferryboat is 2 km from the Muskegon dock when it sounds its foghorn. How many seconds later will people on the dock hear the sound? Give your reasoning and working.

15. *Fireworks*

During a fireworks display, a rocket goes up, then explodes at the top (to make an expanding 'starburst' of light). Why do you see the starburst before hearing the sound? If the sound comes 2.5 seconds after you see the starburst start, how far away did the rocket explode? Give your reasoning/working.

INFERENCES AND PREDICTIONS FROM OUR MODEL OF LIGHT BEHAVIOR

16. *How our abstract model represents behavior of emitted light*

We explored the behavior of light emitted from a point source, and developed an abstract diagrammatic model to represent several of the properties. Sketch the model, and say *which* properties are represented, and *how*.

17. *Light bulb near one end of a room*

A bare light bulb hangs from the ceiling of a square room, closer to one side of the room than the other. What do you expect the wall illumination will be like, on each wall? Compare the walls with each other, and also how the illumination may vary on each wall. Say how the predictions follow from the properties of light represented in our simple mode. Which properties do we use to make this prediction? Then try it out at home to check!

18. *Using the model to predict shadow size behavior (anticipating the next chapter!)*

Draw a point source with a square blocking card near it. Show how lit and dark areas will arise on the wall beyond. If your diagram is to scale, find the length of the shadow on the wall, by construction on the diagram. Then test your predicted shadow size by actual measurement at home or in the lab. Predict what will change if the screen is moved further away. Would the shadow be the same, smaller or larger? Do you see how this prediction follows theoretically from our simple model of light?

19. *Point source and circular hole*

A board with a circular hole in it is placed one third of the way from a point source of light to a screen.

Use our model of light behavior, to predict what will be seen on the screen, and its size in comparison to the hole. (Give as ratio).

EXTENDED SOURCES OF LIGHT

20. Fluorescent tube light and floor illumination

Consider a long fluorescent ceiling light high above the floor. Predict what will be seen on the floor when you switch on the fluorescent tube, and explain your reasoning. Support your answer with reference to some powerful ideas about light. Then try it. Does the floor illumination match your prediction?

Would the result be any different if the fluorescent tube was close to the floor? Explain, with reference to the model. Then try it.

REFLECTING ON KNOWLEDGE: Cognition behind responses

21. *Illumination due to a line source: everyday and scientific modes of thinking*

Consider the question: what will be seen on a screen far away, due to a line shaped light source? A common response is that the illumination on the screen will be the same shape as the source, i.e. a line, though perhaps a bit fuzzy or broadened. When we try it in practice, nature shows us that this prediction is wrong; the screen is uniformly illuminated. But our interest here is the type of *thinking* that leads to the wrong prediction. I.e. *what kinds of ideas and modes of thinking* leads to this common response. Why do you suppose people say this? What modes of thinking might they be using in responding to the question? And what mode of thinking would be desirable? Note that they have lots of experience with fluorescent lights. Discuss all this. Unless we give these issues some attention, and find out how people think, we will not be able to teach as effectively as we could.

Chapter 3

LIGHT, ILLUMINATION AND SEEING

Our everyday experience of light, and the observations we make in studying its behavior, all involve our 'seeing' light, i.e. involve both light and our eyes. Thus our study of light must also involve understanding the role and operation of the eye and how we 'see' light sources and other objects around us.

Note that we can 'see' two different classes of things. Firstly, sources like glowing light bulbs, which we can call *active* light emitters or *primary* sources. Secondly we also see ordinary objects like rocks, trees, household objects, people etc, which are not 'active' emitters of light in the same way as a light bulb – yet somehow we still see them.

The questions of interest are thus: how does it work that we 'see' the active light bulb, and can also 'see' the other objects? What role does our eye play in all this?

We will use an inquiry approach to developing knowledge for this topic, in the sequence

Observe > Wonder > Explore > Ideas > Concepts > Model

We will then apply the knowledge developed to explain new situations and solve problems.

SECTIONS

3-1   How do we 'see' things?

3-2   Common naïve ideas about light and seeing

3-3   Application: The appearance of illuminated objects. The phases of the moon.

PROBLEMS

3-1 HOW DO WE 'SEE' THINGS?

A. INTRODUCTION

Look around you! You can 'see' a variety of things, for instance a glowing light bulb, a candle flame, or ordinary objects like furniture, books, and people. The question arises: how is it that we 'see' all of these things? What is going on, how does the whole process work, and what is the role of light and of the eye?

B. PRIMARY SOURCES – AND HOW WE 'SEE' THEM

The situation

Let us first consider an *active light emitter,* like a light bulb or candle flame. We can also call this a *primary* light source, since it is where the light originates.

Observe the source; clearly your eye is aware of 'seeing' the bulb or flame.

> Picture of source

Focus questions and ideas

We wonder: how does this work? What is happening? And what role is the eye playing? Can we draw a sketch to show what is going on?

Ideas about it.

Think of various ideas and suggest what might be going on.

> Ideas of how this might work

Mechanism and model

The instructor will guide discussion, aiming toward producing a mechanism and model. Then write down your proposed mechanism of how this works, and draw a pictorial model of the process, involving source, light rays and the eye. (Use whiteboards).

> Mechanism of 'seeing' a light bulb or candle

Checking our mechanism and model

If our proposed mechanism of the process is correct, what does this predict about being able to see the bulb if you i. face directly toward it, ii. face at 90° to it, or iii. face away from it?

Then try each. Do these observations support you model of what is going on in seeing an active light emitter or primarily source?

C. SECONDARY SOURCES – AND HOW WE SEE THEM

Introduction

In the previous section we considered primary sources, that emitted light actively of their own accord, and we developed a model of how we 'see' them. However, that most objects around us, e.g. rocks, trees, books, people etc. are *not* active light emitters. A book has no mechanism within it for producing light. How then is it that we can 'see' this kind of object also? And under what conditions?

We need to propose a mechanism of what is going on. Our model will likely involve light, the object and the eye.

Specific example. To work from a specific example, suppose there is a book on the table and a lamp nearby. Your are an observer, as shown in the figure.

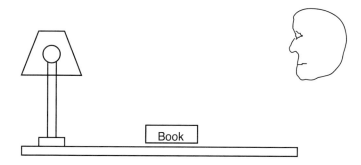

You can certainly see the lamp, an active light emitter, and we have already developed a model for how this works. But in addition you can also see the book. How does this work?

Wondering: What do you think is happening so that you 'see' the book in this setup? Suggest some ideas and models, and illustrate.

Possible ideas:

Idea 1   *Idea 2*   *Idea 3?*

Testing the ideas

How could you check which of these ideas might be viable? Suggest one or more testing experiments and try it out.

Result and conclusion:

Final model and mechanism

Hence propose and draw your (tested) model of how it comes about that we see the book:

Note

Note that we call objects such as the book *illuminated* objects, or *secondary* light sources. According to our model, such an object needs first to be illuminated by another light source, and then it acts as a secondary light source, giving off light from its surface, enabling us to see it.

D. SEEING IN THE DARK?

Focus question

So far, we have developed an idea of how we can see ordinary objects. A natural question arises: what if we are in a completely dark room? Would we be able to see anything?

Imagine you are in a room whose windows and doors are sealed so no light can enter from outside. The room light is initially on, and there is an apple on the table. You turn off the lights; will you be able to see the apple? Why or why not? Explain your thinking.

Your ideas on whether you would be able to see the apple

Then try it, if you have a really well sealed dark room available. What do you discover? How does this fit with your intuitions, and with knowledge from in our previous section on how we see illuminated objects?

Video: peoples' ideas about seeing in the dark

People do have some pre-existing ideas about seeing in a dim or dark room, from experience. These experiences will affect the way we think about seeing in the dark. In a video called "Looking at learning again part 1', students are interviewed about whether they think they will be able to see an apple in the dark.

View the video, then analyze what is happening and what the implications are for teaching about light and seeing. The instructor will guide discussion, and suggest things to look for and think about. These might include: The students' ideas and comparison with scientific ideas; possible origins of students' ideas in prior experience; validity of the ideas in normal situations; current lack of model for reasoning with; implications for instruction, etc.

Discussion and reflections

"Daytime is so you can see where you are going. Nighttime is so you can lie in bed worrying'. – Charlie Brown

E. WHAT HAPPENS WHEN LIGHT STRIKES A SURFACE?

From the previous section, we understand how we 'see' ordinary objects; they are 'illuminated' by light from a source, so that when light encounters the object, some it comes off again and some of this enters the eye.

Focus question

When a light beam strikes an illuminated surface, what happens there?

Exploring

Set up a beam of light (e.g. from a flashlight) and let it strike a matte (non-shiny) surface (e.g. a piece of white paper or the cover of a book). Clearly, you can see the illuminated area of the surface, so light is coming from it to your eye. The question is, after light beams strikes the surface, what happens, i.e. how does it come off again?

Setup

Brainstorm some possible ideas, and sketch a pictorial model of what might be happening at the surface.

Sketches

Idea A	Idea B	More?

Test

Design a test to decide between the competing ideas. Then try it and state the outcome. Which model is consistent with the evidence?

Note on terminology

The process occurring at the matte surface is given the name 'diffuse reflection'.

Notes and reflections

- The situation is a bit different for matte surfaces and smooth shiny surfaces (like a mirror). We investigate mirror surfaces later.

- *Reflecting on learning and naming.* Note that in learning about this we explored the phenomenon first, and only gave it a name (diffuse reflection) at the end. The approach of "idea first, name afterwards" was advocated by Arnold Arons as best for learning new concepts. As soon as people have a 'name' for something there is a danger they will just toss the name around and stop thinking about the basic origin and meaning of the concept!

- Your own reflections on this section

F. CAN WE 'SEE' LIGHT AS IT TRAVELS?

Introduction

An interesting question arises about light and seeing. We know that light from a source will illuminate an object. For example a flashlight illuminating a person at night. If we observe this, we can see the source and see the illumination on the object, but what about the light going in between? We know that light travels, but do we see the path of the light as it travels, or not?

Let's try, using either a flashlight or a laser, in order to get a beam of light. Point it at an object (or at a screen). View this roughly from the side. What do you see?

Can you 'see' any light going between source and screen? Or just the result of it (illumination) striking the object?

Think: Is this observation consistent with our previous ideas about what is required in order for our eyes to 'see' light?

> Setup and observations

Making a light beam visible?

You may have seen movies where the path of laser beams going across a room was very visible. So you may wonder how this fits with our own experiment above, where the beam was not visible from the side.

Any ideas? What could we do to make the path of a light beam 'visible' to us if we looked from the side?

> Ideas?

Remember, we actually have enough knowledge about light and illumination and seeing to work out what must be required. (Or, if you happen to know already how to do it, then you should be able to explain how it works).

The instructor will suggest two simple ways of making the beam path visible. Try both ways, observe the effect, and think about what is actually going on to make this possible. So are we really seeing a traveling light beam from the side, or seeing something else? Illustrate with a sketch, including the eye.

Discussion and explanation:

> Sketch

3-2 COMMON NAÏVE IDEAS ABOUT LIGHT AND SEEING

Identifying common ideas

Seeing things is such a common experience, and our language has various ways of talking about it. Thus we have common naïve ways of thinking about light and seeing, even if this is not always conscious. Many of these conceptions are useful in appropriate situations, others may seem at odds with the scientific view. It is useful to identify and discuss these common ideas and ways of talking, the better to connect them to the scientific ideas, and to discriminate where necessary.

The instructor will lead a discussion.

Discussion

Investigation of people's common ideas and ways of talking about light and seeing

It can be interesting (and illuminating!) to find out how ordinary people, whether adults or children, think and talk about light and seeing. You can do a mini-investigation, preparing a set of demonstrations and questions to ask, and then interviewing a few people. You can report your findings back to the class, and compare with what others discovered.

Implications for learning and instruction

Clearly people's ideas from their everyday experiences on light and seeing will have an effect on learning. So there are implications for instruction. The instructor will lead a discussion.

3-3 APPLICATION:
THE APPEARANCE OF ILLUMINATED OBJECTS

Introduction

Illumination of an object from one direction. When we see various objects around us in everyday life, they are usually illuminated by light from one side more than another.

For example, suppose your friend's face is illuminated by a lamp. If the lamp is off to the side, only that side of her face will be illuminated. If you now look at her from the front, how will her face appear to you? Why?

Of course the lamp may be shining at some other angle to the face; for example it may be placed directly in front of the face, or angled off to the side, or nearly behind the face. In each case a rather different portion of the face will be illuminated, and so the lighting of the face will appear different if you look at her. In everyday life we have long experience of this, we don't notice the different appearances very consciously, and compensate for it when we interpret what we are seeing.

Illumination appearances for a simple shape – a ball lit from one direction

A face is quite a complicated shape, so let's take a simpler one, a ball, and explore illumination appearances carefully. We want to use our knowledge of light behavior to predict how the ball will appear to us when illuminated from different directions.

This makes a good class demonstration. A large ball will held up, and a bright flashlight pointed at it Don't try it yet, but first predict what you will see when the flashlight is pointed at the ball, from different directions, as follows:

Predictions

How would the ball appear to you? Or more accurately, what shape of illumination would you see, when the ball is illuminated from various directions, as below? Sketch the expected appearances.

Beam from front of ball	Beam from back of ball	Beam from side of ball	Beam from front & side	Beam from back & side

Try it. Now try it out and see what you get. Do you see a bright area of the shape you predicted? Do you understand how all this arises?

Remind you of anything? By the way, do the shapes you observe remind you of anything you've seen in ordinary life? What?

Taking illumination appearance ideas further – as far as the moon!

The varying appearance of the moon

We can apply what we have discovered so far to explain the varying appearance of the moon during the month.

Everyone will have noticed that the moon looks very different at different times of the month – it varies from a perfectly round bright shape ('full moon') to a thin crescent shape ('crescent moon') and various shapes in between (each of which has its name). You may remember these shapes, or else can start observing the moon on a daily basis, sketching its appearance over the

Moon appearances

course of a month. (Note that sometimes it will be visible during the night and other times during the day). If you can't wait a month, then calendars often include small sketches of the moon's appearance during the month, and you can use that for now. But it is interesting to make your own moon observations! And of course that is what they did historically.

Sketch the moon shapes or 'phases' below. The phase names are given.

Full moon	'New' moon	Half moon	Gibbous moon	Crescent moon

From ancient times, people observed these changing appearances of the moon and wondered about it. Keep in mind that they didn't actually know what the moon was in those days – they had just their own visual observations to go on, of a bright object in the sky. They noted the changing appearance and how this related to where the sun was, and from this they inferred what was going on and what the actual shape of the moon must be.

So, from your knowledge or observations of moon phases, plus your knowledge of light, illumination and seeing, how would you explain what is going on? What must the actual shape of that object in the sky be, and why does its appearance change?

Furthermore, can you infer what the location of the sun must be, for each of the moon phases in the pictures above?

Optional group work: drawing sun-earth-moon configurations

These days, we know something about the solar system, and how the earth moves around the sun, and the moon around the earth. Using this geometrical perspective, we can draw the relative positions of the sun, earth, observer and moon that would be required to see each moon phase.

The class can cooperate on this in groups using whiteboards. Five groups can each take one of the five phases in the table above. Draw the sun/earth/observer/moon configuration for that phase. The sixth group can physically model each whiteboard diagram as it is presented, using real balls for sun and moon.

Configuration diagrams: (make separate page).

Extension activity

Look at the moon and the sun many times over the course of a month, both day and night, and try to connect the moon's appearance to principles of illumination and seeing, the sun's location, your location, and geometry.

Reflecting on knowledge

While studying the behavior of light and illumination, we have produced some astronomy knowledge! I.e. this was an *application* of pure science knowledge about light to *produce* knowledge about astronomy, from our own observations and ideas. That is, we have *explained* the phases of the moon, in terms of our basic knowledge of light behavior. This is one of the aims of basic science!

Finding diagrams and simulations for the phases of the moon

Find some diagrams, animated simulations and explanations of the phases of the moon, in books or on the Internet. Photocopy and/or download the best ones for teaching and learning. Bring to class. We can look at them, and call up the good URLs on the lab computers.

PROBLEMS

Light, illumination and seeing

Idea power!
Using the knowledge we've developed to solve problems, explain, and predict in new situations.

1. **Seeing a piece of paper**

   Two students have different theories about how we 'see' a piece of white paper. Bob sketches his idea on the left: he proposes that light travels from a lamp to the paper, then 'bounces' off the paper as shown and goes on to enter the eye.

   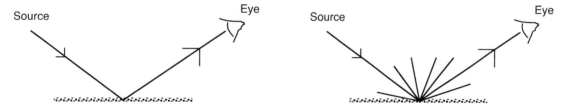

   Sally sketches a different idea on the right: light travels from the lamp, but after it hits the paper, light comes off in all directions from where it hit, and some of this enters the eye. Either model can presumably explain how we see the paper, but we want to know which is better, which can explain more.

   How could they test experimentally whose idea is superior?
   Then try out your own test.

2. **Seeing in the semi-dark**

   Suppose you are in your bedroom at night with the lights on. The lights are then turned off, but a bit of faint light can come in from the window. You can see nothing at first, but after a while you start seeing shapes, like the bed and dresser. What do you suggest is happening here? Why can you see almost nothing at first, then can make out things, but never as clear as when the lights are on? How does this accord with a model of how illumination and seeing works?

3. Photographic fill-in light

In portrait photography, when a face is lit by light partly from one side, you get harsh contrast in the photo, i.e. the lit areas of the face are very bright and the shadow areas very dark. To 'soften' the picture, photographers often place a light-colored umbrella on the opposite side to the lamp, to provide fill-in' light for the shadows. The setup is as shown, seen from the top.

Explain how you think all this might work for getting a good photo. Illustrate with rays involving both the fully lit areas and the 'shadow' areas of the portrait.

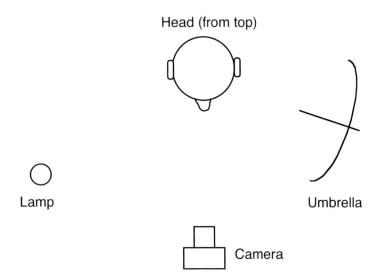

Head (from top)

Lamp

Umbrella

Camera

4. Bounce flash

Photos taken with flash pointed directly at a person often look a bit unnatural because we are used to people's faces being lit partly from above either by the sun or by a room light. To overcome this, one can point the flash at the ceiling instead. Explain how this 'bounce flash' effect works to give more natural lighting. Show the face, ceiling, flash and camera.

5. Bathroom and bedroom

If light travels in straight lines and not around corners, how is it at night that when someone turns on the light in the bathroom down the passage your bedroom becomes a bit lighter? Illustrate what is happening with a sketch. Why do you think the light in your bedroom is so faint in this situation?

6. Why is the shadow not completely black?

In a room, you stand in front of a single small light bulb on the table, and your shadow is cast on the wall. You notice your shadow is not completely black. Why is this – how can it be? Discuss what must be happening, with the aid of a sketch.

'Making the light beam visible'

7. Laser beam

When a laser beam is shone across the room, you cannot see the beam. However, when a water spray is used, you can 'see' the path of the beam. Explain how this works. Are you really seeing the light beam – or what?

8. See light traveling?

If there is light there, between the source and the screen, why can't you see it, if it is indeed there traveling through that space? Better phrased: from the fact that we cannot see it there, what can we infer about light and seeing? Give your best idea. What is required for us to experience 'seeing' light?

9. Headlight beams in the fog

On a clear night, you cannot 'see' a car's headlight beams when you look from the side. Why not? When it is foggy however, you can see a cone of light from each headlight. How does this work? Why is it that when driving in heavy fog at night, it is very hard to see anything ahead in the headlight beam? What do you see instead?

APPEARANCE OF ILLUMINATED SHAPES

10. Cylinder

Lit from the front, side, semi-back or back. How does it appear to the viewer?

11. Moon

Phases

12. Phases of Venus

Photographs of Venus, showing phases. Instructor to provide or refer

What does this evidence tell us about Venus? What shape is Venus? Why does it show these various 'shapes' or phases at different times? Why does its size vary on the different photos, unlike the moon? How does our model of light, illumination and seeing tackle all this?

Chapter 4

MATH INTERLUDE: SOME GEOMETRY AND ALGEBRA USEFUL FOR OPTICS

Straight-line light travel leads to geometrical optics

As long as light travels in straight lines, we can represent light travel by drawing straight *rays*. These rays will form patterns of straight lines in many situations; in particular they will form sides of geometrical figures (such as triangles). Thus our study of light will inherently involve *geometry.* In fact the branch of light dealing with straight rays is called *geometrical optics* or *ray optics.*

Some examples of light topics where the geometry of triangles comes in naturally, are: Shadows, Apertures, Reflection, and Refraction – in fact the whole geometrical optics course!

Geometry and algebra in optics

Since geometrical figures will come into our study of optics topics, it will be useful to refresh our knowledge of triangle geometry, similar triangles in particular. The geometrical properties can also be expressed as *algebraic* relationships, so that algebra will also be useful to our study of light. We can deal with the geometry and algebra either before we start these topics, or later as needed.

It turns out that much the *same* mathematics will apply in various optics topic, even though the physical situations may be very different! This is a powerful unifying feature of math in science, and will become evident in our course.

Goals

Our goals in this chapter will be: to see how geometrical figures arise in ray optics; to study similar triangles and their side ratios; to express these ratios algebraically; and to interpret algebraic relationships in terms of dependencies. In various topic chapters, these mathematical aspects will link very clearly to the physical phenomena.

Notes

This chapter on math for optics can either be studied as background preparation for the topics to come, or it can be skipped for the moment and drawn on at the times needed.

An advantage of introducing the math ourselves in an interlude chapter is we can do it the way we want - we can target it toward optics, and emphasize the useful ways of thinking about it. So even if you already know some of the math this treatment should be useful.

Note also that for purely conceptual (qualitative) courses, many of these mathematical aspects will not be needed, so instructors can use judgment in selecting or omitting parts of the math to suit the nature of their course.

SECTIONS

4-1 HOW GEOMETRICAL FIGURES ARISE IN OPTICS

How triangles feature in ray diagrams – the case of shadow formation

In many physical situations, if we draw straight lines to represent light rays for the system, the set of lines will form parts of geometrical figures, such as triangles.

We can see how such triangles arise in ray optics by taking an example: the case of shadow formation.

The figure shows straight light rays diverging from a point source. Some rays are blocked by a card, resulting in a dark (shadow) region on a screen beyond.

We see that two triangles arise. The sides are formed by the two diverging light rays at the edges of the card, the card, and the shadow. One triangle is bigger than the other, but they have the same shape (i.e. they are 'similar')

The geometry of diverging rays, blocking card, and screen

Note the triangles formed

Thus triangles arise naturally in cases like this. Looking ahead, suppose we want to understand how shadow size might be related to various distances in this situation, viz. distance between source and card, source and screen, and size of card. To relate sizes and distances we will need to understand the geometry of similar triangles, and how the lengths of their sides are related.

In similar fashion, triangles will come into other sections of light, for example apertures, reflection and refraction. This all arises from the fact that light can be represented by straight rays, and these often form sides of triangles in ray diagrams.

Thus, our next task will be to look at the geometry of similar triangles. We will study their properties, seeking relationships between their sides.

4-2 SIMILAR TRIANGLES: THEIR GEOMETRY AND SIDE RATIOS

A. INTRODUCTION

Triangles generally

Triangles come in all shapes and sizes. Draw some of your own, as different as possible.

Various different triangles

Similar triangles

We will be interested in comparing triangles of the *same shape* though different in size. These we call *similar* triangles.

Draw various pairs of similar triangles.

They are *similar* because they have the same shape.

What do we mean by saying that two triangles have the same shape? We mean that corresponding *angles* are the same for the two triangles. Check that this is the case in your diagrams above.

Various pairs of *similar* triangles

B. HOW ARE THE SIDES OF SIMILAR TRIANGLES RELATED?

Focus question

If two triangles are similar, are their sides related to each other in some way?

To study this, let's look at a specific example. The figure shows two similar triangles

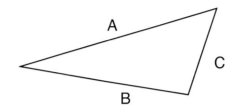 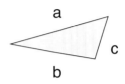

Here the triangles have been drawn so that the corresponding angles are the same, but the sides of the large triangle are clearly longer than the corresponding sides of the small triangle.

Same shape, same angles

How can you *check* that the triangles above are the same shape, i.e. have the same angles? Think a moment. And could you do this without a protractor? How? Ideas?

Of course you could *measure* angles and compare, but a physical way to demonstrate this without measuring is to cut the triangles out and overlap them, as shown. From the figure, it is clear at a glance that corresponding angles are the same. Try it.

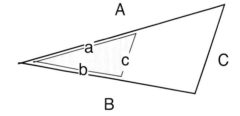

Two similar triangles shown overlapped

Side ratios

By looking at the overlapped triangles above, we can also see how the lengths of sides compare, for this particular example. By measurement, or even by eye, the sides of the large triangle are *twice* as long as those of the small triangle. Check this. That is, check that A is twice a, B is twice b, etc. (With paper cutouts, it is easy to check this by simply moving the cutout small triangle around).

Thus in this example, we see that corresponding sides are in a 2 : 1 (two-to-one) ratio.

We can express this result, in words or symbolically, in a number of equivalent ways, as shown.

Words

– "Side A is twice the length of side a"

– "The ratio of lengths of side A to side a is two to one"

Symbols

– A : a = 2 : 1

– A / a = 2 / 1

Now do the same thing for the other pairs of sides:

In general, one may express the general principle conceptually in words as follows: *For these particular similar triangles, all dimensions of the big triangle are twice those of the small triangle.*

Thus for this pair of triangles, *every* dimension of the large triangle is twice that of the smaller triangle. This means not just the sides, but any other distance dimension. For example, it applies to the perpendicular distance from an apex to the opposite side (often called the 'height' of a triangle), shown dashed in the figures.

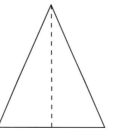

Check it out: measure one of these distances on the large and small triangles shown and calculate the ratio of large to small. Does this accord with the side ratios?

Note that we will sometimes use this feature of a triangle in our treatment of aperture effects.

Of course for other triangle pairs the numerical ratio would be different, but the principle is the same. *Exercise:* Draw a pair of similar triangles where the side ratio is 3 : 1.

Similar triangles with a 3 : 1 side ratio

C. ALTERNATIVE WAY OF LOOKING AT TRIANGLE SIDE RATIOS

There is another way of looking at similar triangles and comparing side ratios. Previously we compared a side of one triangle with the corresponding side of the other and took the ratio, i.e. a ratio *across* triangles. Another way of proceeding is to first work *within* one triangle and take the ratio of two of its own sides; then do the same for the other triangle, again getting a ratio. We then compare the ratios. An example should help make this clear.

Look again at the two similar triangles we used before.

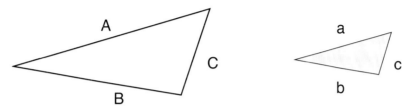

For the *large* triangle, measure sides A and B and take the ratio.

Result for large triangle: A / B = ...

Now for the *small* triangle, measure sides a and b and take the ratio.

Result for small triangle: a / b = ...

What do you notice? . . . .

Let us express this result symbolically:

Ratio A / B (large triangle) = ratio a / b (small triangle)

Or briefly, A / B = a / b.

Notice that we can rearrange this algebraic expression as follows:

If A / B = a / b, then rearranging we get: A / a = B / b.

Do you recognize this last expression?

This last form is exactly how we expressed the side ratios when we compared triangle sides the first way, i.e. *across* triangles! Thus the two approaches are algebraically equivalent – if one form is true then so is the other. They are conceptually and procedurally a bit different however. In work ahead you may prefer to look at a particular case one way or the other, depending on how it best makes sense in the particular problem.

Solving problems using the ratio equation

For similar triangles the sides are related by a side ratio equation:

$$A / a = B / b \qquad\qquad \text{(or } A / B = a / b\text{).}$$

Such an equation is an *algebraic relationship between four quantities* (A, a, B and b). Therefore if we know the value of any *three* of the quantities, we can find the value of the fourth.

We will be doing this for many problems related to light topics in the chapters ahead.

Other pairs of sides

So far for each triangle we considered two of the three sides. In the same way we could write ratio equations between other pairs of sides, i.e. bringing in the third side.

Exercise: Write ratio equations involving all possible pairs of sides.

4-3 EXAMPLE PROBLEMS - Application of knowledge

Instead of just working a set of triangle problems in the abstract, we will use problems which will later have applications to real physical situations. You will not be able to appreciate all the physical consequences at the moment, but it is interesting to know that the math is going to be useful in the real world.

Various configurations of triangles can arise in physics, depending on the phenomenon involved. We provide examples 1, 2, 3 and 4 below. These all involve similar triangles, but the configurations and the way the triangles arise physically in nature are different. Nevertheless, the *mathematical* relationships arising from the similar triangles will be the same for all cases.

Illustrative set of similar triangle problems

We provide a set of four problems which involve the same similar triangles, though in different arrangements. Though the mathematics is common to all, each problem corresponds to a different *physical* situation. Thus two of the triangle problems relate to *shadows*, one involves light going through *apertures*, and one involves *reflection* from mirrors. We will see in later sections how the triangle configurations arise in these different topics.

By working through the problem set now, you can get practice in side ratio problems, and at the same time start to appreciate that the same math can underlie many different physical situations.

Triangle configuration 1

Here is a typical geometry that will arise in the next section, in shadow problems.

The figure shows two overlapping similar triangles, small and large, with some of their sides labeled.

First, identify the similar triangles and say why they are similar.

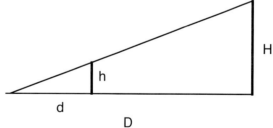

Problem

If the following values are given:

h = 4 cm, d =6 cm, and D = 18 cm, then calculate the length H.

Do this problem in two ways as below. The instructor will give guidance on each method.

  i. Mentally using verbal ratio-logic:

  ii. Formally using algebra:

Physical correspondence.
The *physical* meaning of this geometry diagram will be dealt with in detail in a later section, but in anticipation we state it here. The two diverging lines represent light coming from a point source. The line h represents a card at distance d away, which blocks light and gives rise to a shadow of height H on a screen a distance D away from the source. Thus in calculating the value of length H in the math problem above you have in fact found the height of a shadow on a screen!

Triangle configuration 2

The figure shows a different arrangement of similar triangles; this case arises when two different poles (or trees) cast shadows on the ground. Light rays from the sun are coming down slanted as shown.

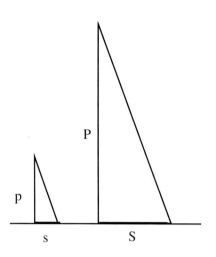

Problem

For these triangles, the following side lengths are given: s = 4 meters, p = 6 meters and P = 18 meters. Calculate the length S.

Do two ways:

    i. Mentally using verbal ratio-logic:

    ii. Formally using algebra:

Physical correspondence. The *physical* meaning of this geometry diagram will be dealt with in detail later, but we give it here for interest. Line p represents a short pole in the ground, and line P a long pole. The shadow cast by the short pole is of length s, and that cast by the long pole is of length S. Thus in solving the problem above you have in fact found the expected length of the tall pole's shadow on the ground!

Triangle configuration 3

When we study the effects of an aperture on light from an extended source, yet another arrangement of similar triangles will arise, as shown in the figure.

Say how you know these triangles are similar:

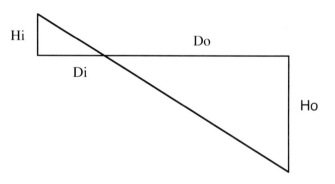

Problem

For these triangles, the following side lengths are given: Hi = 4 cm   Di = 6 cm and Do = 18 cm. Calculate the length Ho.

Do in two ways:

    i. Mentally using verbal ratio-logic:

    ii. Formally using algebra:

Physical correspondence. The *physical* meaning of this diagram will be dealt with later, but we give it here for interest. Line Hi represents the size of an image formed on film in a camera, a distance Di from the camera aperture. Do is the distance from the camera of the object photographed, and Ho is the object size. Thus in solving the problem above we have in fact found the size of an object being photographed, given the image size on film! This may seem mysterious now but will become clear later.

Triangle configuration 4

Yet another arrangement of triangles can arise when a light ray is reflected from a mirror. The diagram shows the similar triangles arising in this situation.

Problem

The following side lengths are given: Dm = 4 feet, Lm = 6 feet and Lj = 18 feet. Calculate the length Dj. Note that the diagram is not drawn to scale, but this doesn't matter for the calculation. Use two methods as before.

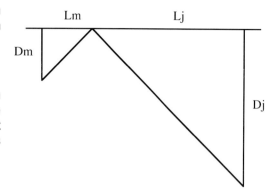

> *Physical correspondence.* The *physical* meaning of this geometry diagram will be dealt with in a later section.

Reflections on all this

Note how the problems are all very similar mathematically, yet they represent quite different physical situations! This shows the generality and power of mathematics, and how, once mastered, it can be drawn on to solve all sorts of problems about nature. Common mathematical threads run through many areas of science! That is one reason we introduce the mathematics early, up front, and give this set of examples anticipating some of the topics that will use the math.

Notice also that we have deliberately chosen a set of examples that are closely related, and with the same numerical values. You will notice that the numerical answers to the different problems are all the same! This is done to emphasize the point that the method is common to all the examples, even though the configurations differ, and the physical situations they represent are completely different!

Notice that we have chosen different symbols for sides in the different examples. When triangles represent physical quantities, it is often helpful to choose symbols that remind you of the physical meaning, e.g. Hi for the height of the image, and Ho for height of the object, and so on.

Your own reflections on the topic and on your own learning:

4-4 ALGEBRAIC EQUATIONS AND DEPENDENCIES

Algebraic equations

Algebraic equations are a way of showing relationships between quantities. For example, the equation

$$y = 3x$$

relates a quantity y to a quantity x. Here x and y are *variables* (i.e. their values can vary), while 3 is a constant (a fixed number). In this case the equation tells us that the value of y is always three times the value of x.

We can look at an algebraic equation in a number of ways, e.g. as a *prescription* or as a *representation*.

As a prescription it is a 'formula' which enables us to calculate the numerical value of one quantity if the values of the other quantities are known. Thus in problem 1 earlier you solved for the shadow height. However note that there was nothing terribly special about the particular given values, and hence nothing special about the particular numerical answer. The problem could well have been set with other values, and we could solve again to get a new answer.

As a representation, an algebraic equation represents a relationship between quantities. It enables us to see the way in which one quantity *depends* on another. Algebra is almost like a language; those who are fluent in it can 'read' an algebraic expression almost at a glance, and see what kind of relationship it represents, i.e. see how one quantity depends on another, which we can call the functional dependence

Even in calculation problems it is useful to work with the algebraic form as long as possible, because it is more general and tells us the functional dependencies, and we can substitute given values toward the end to find the numerical answer.

Example 1

To develop the ideas we need about algebraic equations, it is useful to work with specific examples. First let us look further at the simple equation

$$y = 3x$$

i. 'Value' or 'formula' viewpoint. One way to read this equation is from a 'value' viewpoint: given a particular numerical value of x, the equation lets us calculate the corresponding value of y. Thus if x has value 2 for example, then the value of y will be given by:

$$y = 3x = 3 \times 2 = 6$$

Here we are using the equation as a 'formula' for calculating.

ii. Relationship/dependency viewpoint. The *form* of the equation also tells us the *kind* of relationship between x and y, i.e. it indicates the behavior of y as x changes. In this case, the form of the equation $y = 3x$, (with x and y being in the numerator on each side) tells us a number of things. First of all, it indicates that if x *increases* in value, then y also increases. More precisely, it indicates that if x *doubles* for example, then y will also double. We say that y is *directly proportional* to x.

Physical examples. A physical example would be the relationship between distance covered and travel time. If you travel for twice the time then the distance you cover also doubles, as long as your speed is constant. Distance covered is directly proportional to travel time. Try writing an algebraic equation relating distance to time. Another example we will see later is the relation between shadow size and distance away.

Another way to write the relationship and think about it

The equation y = 3x can also be written in the alternative form

$$\frac{y}{x} = 3$$

One can interpret or 'read' the relationship in this form too, as follows: It directly expresses the fact that the *ratio* of y to x is constant, i.e. the values of x and y can vary, but are related in such a way that dividing y by x always gives the same result. We can also see that if the value of x increases by a certain factor, then the value of y must increase by the same factor (in order for the equation still to hold true). For example if x doubles, y will also double.

Note that saying "the ratio of y to x is constant" expresses the same property as saying that "x and y are directly proportional". You should be able to write the equation in both forms, interpret it both ways, and transform from one form to the other.

Example 2

Now lets take an algebraic equation of a different form, namely

$$y = \frac{6}{x}$$

In this case the variable x is in the denominator.

Lets look at this equation in the same way we did the first, to understand its characteristics.

First take a 'value' or 'formula' viewpoint. For a given value of x, treat this as a formula to get the value of y. Try it, where x has value 3 for example.

Next take a relationship/dependency viewpoint. Lets see what kind of *behavior* the equation represents, i.e. see in what way y *depends* on x, or how y changes when x changes. Here are the questions of interest again:
If the value of x *increases*, what will happen to the value of y? . . (Notice again that x is in the denominator now).
If the value of x *doubles*, what will happen to the value of y?
We say that in this case, y is *inversely proportional* to x.

Physical examples. A physical example of inverse proportionality is the relation between speed and time taken for a given trip. If you travel at twice the speed, you will take half the time to cover a given distance. Another is the relation between volume and

Chapter 5

SHADOWS

– and the geometry and math involved

PROBLEMS

5-1 INTRODUCTION

Shadows

We are all familiar with shadows from everyday experience. But what *are* shadows?

Note that children may start with some different ideas about this than adults.

Own knowledge of light will help us understand shadows and their behavior.

Tasks:

i. Ask some young children for their ideas about shadows and how they come about.

ii. Read up some work on what children think about shadows.

Shadows fun

Just for fun, we start with some amusing examples of shadows and shadow behavior, some of which you may have already tried when younger. These demonstrations and activities can be used in classrooms as a fun introduction to a serious study of shadows.

- Making shadow animals on the wall.

- Shadow theater.

- Shadow poems.

- Try to beat your shadow.

-

Shadow poems

Write your own shadow poem, and find an existing one that you like.

The instructor has one to share also

The instructor will introduce, and guide discussion.

5-2 HOW SHADOWS ARISE – exploring and explaining

It is useful to start thinking (and teaching) about shadows by observing the real thing.

First set up a point source of light, so that it illuminates a screen. Then put square card in between and see what happens! Describe.

Focus question

What is this 'shadow'? How does it come about?

From observations with this arrangement, people can usually suggest what is going on, even with no formal science knowledge.

Explanation in words of how shadows arise

Of course we have the advantage of having already developed a model of light behavior, so we can use it to explain shadow formation.

A side view of the setup is shown on the left of the figure. On the right we have added a 'front' view of the screen, to show what the shadow looks like viewed from the front.

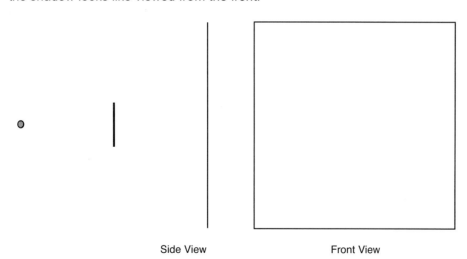

Side View   Front View

We can use our model of light behavior to explain how the shadow comes about. This will include showing how light is emitted from the source, and using many rays, to show how 'lit' and 'unlit' regions arise on the screen.

Add rays to the diagram above to do the following:

- Use the model to represent how light is emitted from the source.

- Using many rays, show how 'lit' regions arise on the screen (in side view).

- Using many rays, show why there will be a 'dark' region on the screen (in side view).

- How will the light intensity on the lit areas of the screen compare with the light intensity on the blocking card? More intense, less intense or the same? Explain how this is shown by the model.

- How about the size of the shadow? Will it be larger, smaller or the same size as the blocking card? (Note how our model is a 'natural' in showing why this will be the case).

- Compare the height of the shadow with the height of the blocking card. What is their ratio, for this particular setup?

- Does the model predict the *shape* of the shadow also? Draw in the predicted shape on the screen front view.

Testing by checking this out in practice

We can test the above experimentally, and see if nature agrees with our model predictions! Set up a blocker card midway between a source and a screen. Check if the shape and size of the shadow is as you expect, and check how the intensities of light on the card and screen compare. Report, and say if the model's predictions are confirmed.

If this all checks out, then we have some confidence in the power of our model.

5-3 SHADOWS SCHEMA 1

This is a good time to look back on what we have learned about shadow formation so far, and devise a 'topic knowledge subassembly' for it. This is likely to be just a 'phenomenon diagram' at this early stage – other subassembly diagrams will come later.

Phenomenon diagram for shadow formation

It is useful to have a mental diagram that contains the essence of how a shadow is formed. It would summarize the phenomenon in a simple abstract way, and could act to 'cue' the knowledge we have developed when we need it.

In this case, the phenomenon diagram might be a simple sketch of the shadow formation diagram you developed in the preceding section. Identify the essential aspects you want to show, and list them:

Check you list against these suggested aspects: light source; emitted rays; blocker; some rays blocked and others not; lit and dark areas on a screen. (Note we include both the lit and dark areas, both part of the phenomenon of shadows).

Sketch your phenomenon diagram:

5-4 SHADOW BEHAVIOR – conceptual approach

If the above was all there was to understanding shadows (knowing how they arise) we could stop right here! But while exploring shadows a bit, you probably noticed that as you moved the source, card and screen about, the size of the shadow changed. So, we would also like to understand this shadow *behavior*. And not only this, but also be able to work out exactly what shadow size will be produced for a given setup.

Thus his section will be about applying our model of light behavior to explain shadow behavior and solve problems. It simultaneously involves the mathematics needed for dealing with shadow situations – namely the geometry of triangles and the ratios of their sides.

Introduction and exploration

It is useful to approach a topic *qualitatively* first, to get a conceptual understanding of the system and its behavior, before quantitative measurements or calculations. That is, we deal with the *concepts* and *type of behavior* first, before bringing in math, formulas, numbers and calculations. The quantitative aspects are left to a later section.

We thus start this section by exploring shadow behavior *qualitatively*, i.e. exploring what *kind* of behavior occurs, rather than trying to calculate numerical values, at this stage.

Shadow setup

Our first simple shadow setup will be a point source, a blocking card and a screen. Set it up, and sketch it.

We can now make various *changes* to this setup, e.g. we can move the screen, the source or the card, thereby changing the distances involved, and see what happens. We can also change the size of the card. We will explore how shadows *behave* as we make these changes.

Sketch of setup

Focus question

As we adjust the positions of screen, source or card, what happens to the shadow?

Exploring

Observe the effects of moving each of these objects, one at a time. See how the shadow size changes in each case, and note your observations.

Observations
Effect of moving screen further away:
Effect of moving source further away:
Effect of moving card closer to source:
Effect of a larger card:

Explaining the effects

Our aim will now be to *explain* the behavior, in terms of our *model* of light. Do these shadow effects follow from the powerful ideas about light, as reflected in our model? Let's apply the model to four cases. Do it on whiteboards first, then transfer your diagrams to the space below.

Case 1: Effect of moving the screen further back. Make a whiteboard diagram to explain what happens to the shadow.

Case 2: Effect of moving the source further away. Make a diagram to explain what happens to the shadow.

Case 3: Effect of moving the card toward the source. Make a diagram to explain what happens to the shadow.

Case 4. Effect of increasing the card size. Make a diagram to explain what happens to the shadow.

Explanatory diagrams

Shadow width

We have concentrated on shadow *height*. What happens to the shadow *width*? Does it also vary in the same way or not? Be able to explain.

5-5 SHADOWS SCHEMA 2

Now that we have studied shadow **behavior**, i.e.how shadows change as various aspects of the situation are **varied**, it is a good time to sketch another 'topic knowledge subassembly'. This time it can be a combined diagram, a principles and processes diagram also serving as a 'behavior diagram'.

A principles/processes/behavior diagram for shadows

What 'knowledge sub-assembly' can you devise to show (abstractly) how shadows arise *and* how shadow size varies with distances? Think for a moment.

Even though this ray diagram is static, you can still use it to visualize what would happen if various things were changed in the setup. Thus for example, imagine that the screen being moved further away; it is easy to imagine the diverging rays going on further, striking wider on the screen, and hence we see that the shadow would become larger.

Notes

Reflecting on knowledge gained

We realize that here is nothing to 'commit to memory' about how shadows behave in these various situations, as you adjust one thing or another, since one can always work it out mentally in a flash! The shadow behavior follows directly from the fact that light travels out from a source in straight lines in all directions!

Other shadow geometries

The case we have looked at, shadows from a card parallel to a screen, is reasonably simple and hence useful for exploring shadow behavior and developing the math that goes with it. So we will use it a lot! But of course there are other shadow geometries. In the next section we look briefly at shadows formed on the ground.

5-6 OTHER SHADOW GEOMETRIES

Shadows on the ground

A common shadow geometry is that of shadows formed on the ground, with the light source being either the sun or a lamp high above. In the familiar case of your own shadow, you are vertical and your ground shadow is horizontal.

Sketch the situation where you are standing in the vicinity of a street lamp at night. Use our model to show how the ground shadow arises.

How your ground shadow arises

How your ground shadow varies

For this street lamp example, here are some qualitative questions. Illustrate your answers with sketches.

i. Where do you stand to get the *minimum* shadow size?

ii. What happens to your shadow length as you move further from the lamp?

III. Under what conditions is your shadow exactly as long as you are tall?

How your ground shadow varies

Try it out

Now try it out, using a lamp and a vertical stick, to form a shadow on the tabletop or floor, and see how it behaves.

Your shadow in sunlight

Your shadow in *sunlight* is a similar situation, except that the sun is so far away that all rays reaching you can be taken as nearly *parallel*, rather than diverging as for the street lamp (Ask the instructor about this).

How does your shadow behave as the sun rises from the east, climbs higher, reaches a maximum elevation about the middle of the day, then gets lower again toward the west, and finally sets? Explain how your shadow will behave during this process – and check it out the next convenient sunny day!

Notes

5-7 SHADOWS USING SIMULATIONS

What is a simulation?

> *Simulation:* A representation of a system by a device
> or diagram that imitates the behavior of the system.

Simulations can represent system behavior

We can explain behavior and solve problems in a number of
ways. One way is by *simulation* of the system. By this we mean
that we *model* the system in some way, and the model behaves in
many ways like the real system. Then we can study how the
simulation behaves – and even make measurements – to model
the behavior of the real thing

For the case of light, an important property to be modeled is that
light travels in straight lines. We can simulate this behavior in a
number of ways:

   a) Diagrammatically, by drawing straight *lines* (rays) on paper.

   b) Physically by using *strings* to model light rays.

   c) By using a *computer* to draw rays on the screen.

Note that simulations can be used to deal with both *qualitative*
and *quantitative* problems. Qualitative simulations show what
kinds of things happen, and how things change as you make
adjustments. Quantitative simulations can be used to get
numerical values.

Qualitative simulation

Section 5-3 already contains simulation by ray diagrams of
qualitative shadow behavior. We just did not call it simulation at
the time. The ray diagrams essentially model the behavior of the
actual system, for various situations. Review that section.

Quantitative simulation – to solve a problem.

Let us take a numerical problem, to be solved quantitatively by
simulation methods.

> *Problem.* A screen is 90 cm from a point source, and we have
> a blocker card 20 cm high. The problem is: *where must we
> place the card to get a shadow exactly 30 cm high on the
> screen?* Or to put it another way, *how many cm from the
> source must we place the card?* (Do this problem by
> simulation rather than calculation).

We will get a numerical answer to this by *simulating* light
behavior, either by accurate construction on a scale diagram or by
making a string model. Let's do it both ways.

a) Simulation by drawing a scale diagram

Do the diagram full-scale on a large whiteboard. Show your construction and answer.

You will also need to make a smaller scale drawing on paper to go into your notebook,

b) Simulation by strings

Do this in the lab or at home, and give your method and answer.

Checking your answer against the real system – nature

Then check it all out by seeing if nature gives the same answer! Use a real source, card and screen, and see if it gives the shadow size that your simulation predicted.

Notes

I.   Notice how simulations provide a visual *conceptual* understanding of the system and its behavior.

II.   String simulations are fun for young children, and useful for visualizing abstract 'rays' of light.

III.   String simulations are useful for showing three-dimensional situations. Paper only has two dimensions so we have to imagine the third or show it as perspective.

5-8 SHADOW PROBLEMS USING 'VERBAL RATIO-LOGIC'

Introduction

Shadow problems usually involve the side ratios of similar triangles. Here we will tackle shadow problems using 'ratio logic'; that is, we express the steps in the solution as 'verbal logic' statements rather than algebraic formulas. The algebraic approach is left to the next section. Expressing the logic of what we are doing verbally helps promote conceptual understanding. The method is particularly transparent with simple whole number data so we can use mental arithmetic for calculations.

A good way to illustrate the approach is to work through some particular examples.

Example 1. Getting height ratio from distance ratio

A card is between a point source and a screen, so that a shadow of the card is cast on the screen.

Suppose the screen is twice as far as the card from the source. How many times taller will the shadow be than the card? Or putting it another way, what will be the ratio of the shadow height to the card height?

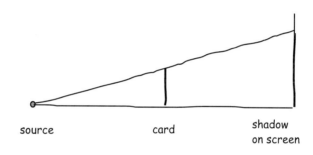

source   card   shadow
                              on screen

Solve this problem by ratio-logic, guided as follows:

> The screen distance from the source is . . . . . times the card distance from the source.
> Therefore, the shadow height on the screen will be . . . . . times the card height.
>
> Alternative wording of the same reasoning but using the term 'ratio' would be:
>
> The ratio of screen distance to card distance is . . . . . .
> Therefore the ratio of shadow height to card height will be . . . . . .

You can set up the actual equipment to check your answer against nature if you like.

Note
Only a *ratio* is asked for in this problem, rather than the actual size of the shadow. Other problems might ask for the latter, which would require another step in the logic.

Example 2. Shadow height using ratio logic method

A card of height 12 cm is 20 cm from a point source and a screen is 60 cm from the source.
What will be the height of the card's shadow on the screen?

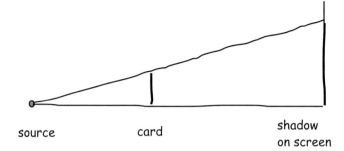

source   card   shadow
                              on screen

Solve this problem by ratio logic, guided as follows:

> The screen distance is . . . . . .cm and the card distance is . . . . . cm.
> Therefore the ratio of screen distance to card distance is . . . .
> Therefore the ratio of shadow height to card height will also be . . . . . .
> The height of the card is . . . cm, so the height of the shadow will be . . . . times this,
> i.e. height will be . . . . .cm.

Reality check

You can set up the actual equipment to check your answer against nature.

5-9 SHADOWS USING MATHEMATICS

Introduction – math and science

This section illustrates another aspect of physics – how it uses mathematics as a language for describing the behavior of the physical world! We will see how "Math Power" can be incredibly useful in science. Mathematics can be seen as a tool, a representation, a way of thinking, a concise language. It can *represent* the properties of real systems in an elegant and powerful way. Math and science are thus partners. Developing proficiency in math and being able to interpret it physically is important for understanding science.

For shadow problems, what math will we require? The geometry of similar triangles, the resulting ratio relationships, and some basic algebra.

The use and power of mathematics in physics is illustrated by the following problem on shadows.

Example problem 1

> **Problem 1. How tall a shadow?**
>
> You have a small light bulb in a socket, a square card in a holder, and a screen on a stand.
>
> The card is 2 ft high and is 4 ft from the bulb, while the screen is 6 ft from the bulb. What will be the height H_s of the shadow on the screen?
>
> Solve this problem *mathematically*, using similar triangles and ratio relationships.
> Use algebraic symbols for the quantities involved, only substituting values at the end.

Solution procedure

We will go carefully through **all** the stages of representing and solving this problem.

a) Reality sketch

A useful first step is to sketch the setup – i.e. draw a 'reality picture'. This translates the *words* of the problem into a visual picture (which you would have to imagine anyway even if you didn't sketch it).

Sketch of setup, with given values

b) Abstract diagram to represent the problem

Next we represent the problem by an **abstract** diagram. This shows only the *essence* of the problem, those aspects needed for its solution. It does not show irrelevant details such as the card holder or what the light bulb looks like, etc. It adds abstractions such as lines for certain light rays, to show the geometry involved.

The abstract diagram can be freehand, and not to scale, since a feature of the mathematical method is that we get the result by *calculation*, not by accurate construction.

Draw a freehand abstract diagram for this problem. Label relevant quantities with suitable *symbols*. Note that you can always choose what symbols to use, but here we suggest the subscripted symbols H_c D_c, H_s, and D_s. In this notation H is used to indicate a <u>H</u>eight, and D a <u>D</u>istance, while the subscripts b and s refer to the <u>c</u>ard and <u>s</u>creen respectively.

> Abstract diagram of the essence only, labeled with symbols

c) Algebraic relationship

From the geometry in your abstract diagram, set up an algebraic equation relating the sides of the two triangles (ratio equation).

> Algebraic relationship

d) Rearranging, substituting values and calculating

Rearrange the equation if necessary to get an expression for the quantity you want, then substitute numbers and calculate the numerical answer.

> Rearrange, substitute, calculate

e) Result:

Note

Work algebraically first, substituting later

Note that we worked *algebraically* with symbols as long as possible, and only substituted numerical values toward the end of the problem. This is good practice. The algebraic form is more general, and applies whatever the particular given values might be. It also makes the nature of the relationship clear.

Comparison with scale construction (simulation) method

It is of interest to solve the same problem by constructing a scale diagram, and see if the construction method gives the same result as the math calculation. Let's try.

Construction to scale

Result

Do the math and construction methods give the same result?

Aim to become expert at all methods!

PROBLEM 2 (Variation on the same theme)

We take exactly the same physical situation as in Example 1 above, but make a variation in the problem. This time the shadow size is specified, and we ask what screen distance is required to get this shadow size. Thus the modified problem is as follows:

Problem 2. Calculate where to put the screen

You have a light bulb in a socket, a square card in a holder, and a screen on a stand.

The card is 2 ft high and is 4 ft from the bulb, and want to place a screen so that the shadow cast on it is 3 ft high. Where must you place the screen?

Solve this problem *mathematically*, using similar triangles and ratio relationships.
Use algebraic symbols for the quantities involved, only substituting values at the end.

Solve this problem mathematically, using the same procedures as for Example 1.

Problem 3: X marks the spot – a prediction challenge

Here is another variation on the problem, posed as a practical prediction challenge!

This is how it works. Lay a large whiteboard on your bench, and have available a maglite source and a screen (but no blocking card). The instructor will place the maglite and the screen on the whiteboard at places of her own choosing (see figure). You should mark these locations on the whiteboard. The instructor also carries a special blocking card, but for later use.

Where to put the blocker?

Challenge

Your challenge is to predict *mathematically* (using geometry and ratio calculations) where the instructor's blocking card should be placed to get a shadow on the screen exactly **two-and-a-half** times the card size.

Math calculation

Solve the problem using math. Then mark your predicted spot with an X on the whiteboard. Show your calculations and constructions on the whiteboard also, for the instructor to look at.

Instructor's test of your prediction

Call the instructor, who will then place her special blocker card at your predicted spot X. This will test your prediction against reality – is the shadow indeed two-and-a-half times the card size?

Your group may be called on to show your whiteboard to the class and talk about your method.

Keep a record in your notebook of your group's work, including diagram, data, method, results and discussion.

Trying your own variations on a theme

For the same physical shadow situation, we can set more problem variants, depending on which quantities we specify and which we want to solve for. Make up one or two variants for yourself, and solve them, until you are confident.

5-10 DEPENDENCIES

– how one quantity depends on another

Recap: how algebraic equations show dependencies

Algebraic equations are useful in a number of ways. They can enable us to solve for the value of one of the quantities, if the others are known. But more than that, they enable us to see the way in which any one quantity *depends* on another! That is, the form of the equation shows *dependencies* between quantities. It thus makes evident the *behavior* of the system it represents.   In our case, we have already set up algebraric equations for shadows, so we should be able to see how shadow heights and distances relate to each other, and hence what will happen to one quantity if we change another. For example, the equation should show us what will happen to shadow height if we change screen distance from the source.

Dependencies are discussed in detail in Chapter 4. You should review that as mathematical background to this section.

Application to ratio equations for similar triangles

Our side-ratio equations for similar triangles have a form which indicates direct proportion between some quantities and inverse proportion between others

 For example, here is an equation for shadow height from an earlier problem (problem 1)

$$H_S = \frac{D_S \times H_C}{D_C}$$

From the form of this equation, say how shadow height H_S will vary if you increase the shadow distance Ds, the card distance Dc, or the card height Hc.

EXAMPLE PROBLEM ON DEPENDENCIES

- Dependency of shadow size on screen distance

Consider the following qualitative question, expressed in terms of dependency:

> A card placed between a point source and a screen casts a shadow on the screen. What will happen to the shadow size if you move the screen further back? Or to put it another way, how does shadow size depend on screen distance?

We will answer this by inspection of the form of the algebraic relationship between the quantities involved. First however we must obtain the relationship, for this situation. We start with a diagram of the geometry:

Draw rough sketch of setup:

Hence produce algebraic equation for the situation:

Look at your equation, focusing in particular on the two quantities of interest in the question, i.e. screen distance D_s and shadow size H_s. Now imagine that D_s is increased, and see how H_s must change, in order to keep the equation true. (Note that if one side of an equation increases the other side must increase likewise, for both sides to remain equal).

Your answer and justification:

Note

From the form of the equation, inspecting how the quantities D_S and H_S appear in it, we can 'see' that D_S is directly proportional to H_S. (Both quantities are in the numerator, on opposite sides of the equation). Symbolically we can write this proportionality as

$$D_S \propto H_S.$$

The proportionality means that if D_S is changed then H_S changes in exactly the same way. This all assumes that the other quantities in the equation are kept fixed.

Checking this dependency in another way – by geometry rather than algebra

We can check this dependency in another way by drawing a suitable ray diagram. Then by looking at the ray geometry on the diagram we can see what would happen if we move the screen back. Try it. Is your conclusion consistent with what your algebraic equation indicated?

Checking against nature: behavior of the real thing

Finally, we can see whether the behavior indicated by our simulations and representations is what actually happens with real shadows. Set up the shadow situation, vary quantities, and see what happens. Is this consistent with our 'theoretical' predictions?

Note

This section has shown the power of algebraic representations! Algebra can represent relationships and physical systems very elegantly. Physics and math work very closely together! Physics uses math as part of its 'language' in representing the natural world and its behavior.

Simple *quantitative* dependencies

So far the dependency discussion has been pure qualitative: we have only asked about a quantity *increasing* or *decreasing*. But the equation tells us more than that: it tells *exactly* how one quantity depends on another, i.e. it gives quantitative dependencies too. A way to illustrate how this works is to let *one* quantity change by a certain amount, like a factor of 2 for example (a simple whole number for simplicity), and see by what factor the *other* quantity changes. So, in our example, imagine **doubling** the screen distance. By what factor will the shadow height change?

106

Dependencies involving other quantities in this shadow situation

Note that the algebraic expression you used for this problem is actually a relation between **four** quantities, H_b, D_b, H_s and D_s. So far we have used it to see how H_s depends on D_s, with the other two quantities taken as fixed. That is, we have chosen to view D_s as the 'independent' variable and H_s as the 'dependent' variable, with the other two quantities kept constant. But the algebraic relation gives *all* the interdependencies between the four quantities in the situation. In posing a problem we can decide what quantities to fix and what quantities to allow to vary.

As an example, let us look at all three quantities that the shadow height H_s depends on: it depends on the screen distance D_s, as we have seen before, and it also depends on the blocker height H_b and the blocker distance D_b. Let us now consider these last two:

1. How shadow height depends on blocker height

• Start with a sketch. • Show *algebraically* how shadow height H_s depends on blocker height H_b. • Then consider whether your answer makes *physical* sense, referring to a ray sketch. • Then finally try it out for *real*; as you increase one quantity, see what really happens to the other one, and say whether this agrees with the dependency reflected in the equation.

2. How shadow height depends on blocker distance

• Start with a sketch • Develop the algebraic relationship from the sketch • Show *algebraically* how shadow height H_s depends on blocker distance D_b. • Then consider whether your answer makes *physical* sense, referring to a ray sketch. • Then finally try it out for *real*; as you increase one quantity, see what really happens to the other one, and say whether this agrees with the dependency reflected in the equation.

Challenge:

We can also ask the following: if we move the blocker, is there a *maximum* shadow height we can get, and what setup will give this? See if you can answer this, using both the physical diagram and the algebraic relation!

Homework practice: further possible variants of this shadow problem

One can of course construct further variants of the shadow dependency problem, choosing other dependent and independent variables, and fixing the other quantities. For example, one could ask how required blocker distance depends on desired shadow size, or how required screen distance depends on blocker distance etc. Try this.

Reflections and insights on math and shadows

Algebra power

The power of mathematics became evident in this shadow situation! We saw how math expressions can represent the behavior of real physical systems, and can do so efficiently and compactly. That's a reason why math is part of the 'language' of physics. As we become proficient in interpreting math equations and relating them to real physical situations, we are becoming fluent in the language!

Note also how algebraic equations can be used to answer both qualitative and quantitative problems.

Geometry power

We have also seen the power of scale construction methods, which are a graphic simulation of light's actual behavior. They represent another side of math, geometric rather than algebraic. This geometric approach is conceptually helpful in understanding physical systems.

Note that geometry too can be used for both qualitative and quantitative problems.

Become a well-rounded expert

Aim to become expert in both geometrical and algebraic methods!

Your own reflections and class reflections

5-11 SHADOWS ON THE GROUND

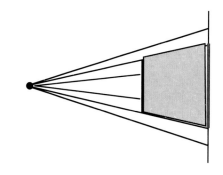

Shadows so far

So far, we have mostly looked at one type of shadow geometry, namely 'shadows-on-the-wall'. That is, we had a light, a blocker card, and a screen all lined up, with the card parallel to the wall, as shown in the figure alongside. This was a useful geometry for developing most of our important insights about shadow behavior and the related mathematics. We now look at other shadow geometries.

Shadows on the ground

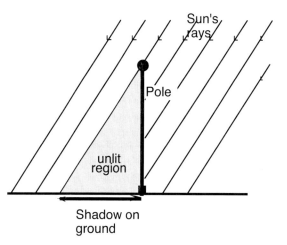

Shadows occur in all sorts of situations. One case familiar from outdoor life is shadows-on-the-ground. Here the sun is the source of light, and things like people, trees and poles cast shadows on the ground. The figure alongside shows the geometry, for a vertical pole casting a horizontal shadow on the ground. Note how the light-blocker (pole) is at right angles to the surface (ground) on which the shadow is cast.

Note too that the sun's rays are drawn as coming in *parallel*, not diverging. Is this correct, incorrect, or nearly correct? Why? The sun is so far away that all rays reaching you can be taken as virtually *parallel*, or very nearly so (rather than diverging). (Ask the instructor about this).

Basic mechanism of shadow formation

Despite different configurations, you will see that the *basic mechanism of shadow formation* is the same for all cases. Right? An object blocks some of the light rays. Certainly the geometry is different between cases, but the physics concepts apply just as before, and the same math will be useful in problems.

There are interesting examples and problems for shadow-on-the-ground situations, which we turn to next

Problem 1 - Your own shadow during the day

How does your shadow behave as the sun rises from the east, then climbs higher, reaching a maximum elevation about the middle of the day, then gets lower again westward, then finally sets? Explain how your shadow will behave during this process and why. Illustrate with sketches, and check it out the next convenient sunny day!

Problem 2 – finding the height of a tall flagpole or tree

Physics has useful practical applications. Suppose you come across a very tall flagpole (or tree or skyscraper), and want to know its height. How could you determine this? Can you think of a clever way to work it out without actually having to go up the flagpole? Can you use some of your physics knowledge plus some creative thinking? Think and discuss before reading on.

One idea is to use a smaller scale model – e.g. a vertical stick as a shorter comparison to the flagpole – and compare the shadows formed by both in the sunlight. To solve the problem we will need to know some physics (how shadows are formed) plus some math (geometry and ratios). The diagram shows the situation.

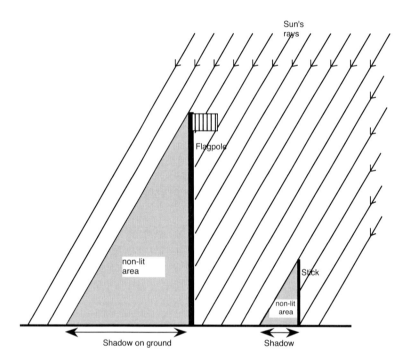

The sun's rays come in at an angle and form a shadow of the flagpole on the ground as shown. Nearby, you can hold the stick vertically, and this likewise forms a shadow (a shorter one as you see). The stick with its shadow form a small-scale model of the flagpole situation, one that we can more easily measure.

By now you may already see how to tackle the problem of finding flagpole height! But first, to appreciate the geometry of the situation fully, let's analyze the properties of the diagram, as follows:

Sun's rays

Why can we draw the sun's rays as essentially parallel in this situation, instead of diverging?

Triangles

Notice how the 'shadow triangles' for the flagpole and for the stick are similar to each other.

110

Ratios

Measure the heights of the flagpole and the stick on the diagram

Find the ratio of flagpole height to stick height. . . . .

Measure the lengths of their shadows . . . .

Find the length ratio of the flagpole shadow to the stick shadow. .

Compare the height ratio with the shadow ratio. Notice anything?

Note: Thus you can see why ratios are an important part of this problem situation!

An alternative set of ratios

Lets look at another way to take ratios in this situation:

Find the ratio of the height of the flagpole to its shadow length.

Find the ratio of stick height to its shadow length.

Compare. Notice anything?

Note: So, there is more than one way to look at ratios in this situation. Learn to use both!

Back to the problem of finding flagpole height

From the diagram geometry, can you suggest how to find the flagpole height, if you measure the lengths of both shadows, and know the height of the stick? Describe your method below:

Your proposed method:

Finally – do this for a real flagpole, outdoors!

We have worked out how to use shadows, similar triangles and ratios to find the height of a tall object (flagpole, tree, tower, or building). Now let's actually do it in practice.

As an added challenge, let's say you have no measuring rule available. (This is likely to be the case if you were really out in the wild and someone asked you about the height of a tall tree). So you have to decide how to measure distances, and what units you will use (or invent) for doing this. Remember, anything can serve as your group's own chosen length unit! (Of course you can later refer your own length unit to conventional units if you wish).

And what if there was no stick available to hold vertically? What might you use instead? Ideas?

Proposed procedure

The method you used, your units, your data and your results:
Present your groups' method and results to the class. How does it compare with other groups?

Bad weather alternative activity – modeling the situation

If the weather is cloudy so there are no shadows, we can set up a smalll-scale *model* in the lab to show the situation Place a small light bulb high up in the corner of the room to represent the sun. Stand a tall object on a horizontal whiteboard quite far from the bulb, and note it casts a shadow. Then stand a reference object next to it on the whiteboard. By comparing shadow lengths you know how the heights relate, and hence can calculate the height of the tall object (representing a flagpole or tree)

Reflections, think-back, debriefing

- What was the essence of this section? Summarize the main thrust for yourself.

- Note how the physics and math knowledge we developed for light turns out to have practical uses in real life situations!

- Your own reflections: (n the topic and on your learning:

5-12 SHADOWS FOR MORE THAN ONE SOURCE
– what kinds of shadows do we get?

Introduction

So far, we have investigated shadows arising from just a *single point source*. Now, we ask: what if there are *two* sources? What sort of shadows will arise and why? And if the light source is an *extended source*, not a point source? What then?

We will treat these more complex situations as application challenge problems. Using 'idea power' we will tackle them from basics using our few powerful ideas.

Shadows due to two or more sources

Focus question

Consider a situation where there are **two** point sources of light, with a blocking card nearby. What would be seen on a screen beyond? Any initial ideas?

Picture

Approaches

There are two ways to approach this: either as a *prediction* challenge or an *explanation* challenge.

i. Prediction challenge. Here we first **predict** what will happen, by applying what we know about light, and then try it out in practice to see if our prediction is confirmed. That is, we go from theory to experiment.

ii. Explanation challenge. Here we first **observe** what happens, then apply our knowledge of light to **explain** what we observed. That is, we go from experiment to theory.

We will try these approaches in the cases below.

Cases

For shadows from two sources, we will find somewhat different effects depending on whether the sources are far apart or closer together, relative to their distance to the card. (Sketch)

The two cases are given below. Treat one of them as a prediction challenge, and the other as an explanation challenge!

a)   Two sources relatively far apart.

Are you treating this theoretically or practically first?

Include observations, ray diagram and explanation.

Ray diagram (a)

b)   Two sources relatively close together.

Include observations, ray diagram and explanation.

Ray diagram (b)

c)   Going from one case to the other.

Start with the sources far apart and move them slowly closer together, noting what happens. State your observations. Does this behavior make sense?

Reflections: insights on this

Would you agree that shadows from two sources are just the 'combined effect' of the shadowing process from each separate source? So our previous few powerful ideas handled this just fine! We needed no new physics to deal with this more complex situation. Thus an aim of science is realized – idea power!

Shadows due to *multiple* sources

Once we understand the principles behind shadow formation, and can apply our model of light, we can readily tackle the case of shadows due to three or more point sources. There is no new physics involved, so we can simple treat any such case as an exercise in applying the principles. Since this is application, we can leave it to the problem section.

5-13 A MYSTERY PRELUDE- FUZZY SHADOWS!
- the edges of your shadow in the sunlight

When you are outside in the sunlight, have you ever looked closely at your shadow on the ground? What about its edges? Do you recall if your shadow has sharp or fuzzy edges?

Well, go outside and look! What do you discover? Describe.

This is a mystery! From what you know about light behavior so far, would you have expected this?

Are you puzzled? Does it make sense? Do we have to go back to the drawing board on light and shadow, or can we explain it from what we already know?

Take it as a mystery to be solved. What could be going on here?

Think, discuss in groups and try to figure it out. Use knowledge, experiment and thought. List any ideas your group comes up with.

If we have ideas about this, how could we *test* them? What could we do? Either by experimenting outside in the sunlight, or by making a mock-up of the situation and trying it in the lab. Instructor will lead a discussion.

5-14 SHADOWS FROM E X T E N D E D SOURCES OF LIGHT
– what will shadows be like if the light source has 'extent'?

Earlier we looked at shadows from a **single** point source of light. Then we considered shadows produced by **two** (or more) point sources, and got some interesting effects. The obvious next question is: what about shadows from an **extended** source of light?

Situation:

Given an extended source of light in the shape of a **line**, a rectangular blocker card, and a screen.

We wonder: for this line source, what will the shadow be like, and why? Will our existing powerful ideas about light be able to explain what we get?

Picture: extended line source, square card, and screen

Idea power – predict what will be seen on the screen

Let's put our physics 'idea power' to work on this new challenge! Treat it as a **prediction** challenge, that is, apply your theoretical model about light from extended sources to work out theoretically what should happen, before actually trying it in practice.

Using the model and ray constructions, figure out what will be seen on the screen.

Ray diagram

Prediction:

Then try it in practice

What do you observe? Was your prediction correct?

5-15 REFLECTIONS

Powerful ideas

Using a few basic 'powerful ideas', viz. that light travels in straight lines, is emitted in all directions from a point source, and that an extended source is just a collection of point sources, we were able to explain shadows and their behavior, in a wide range of situations. This highlights just how powerful and versatile those few 'powerful ideas' are! That is an aim of science.

Powerful models

Our ray model of light behavior incorporated those powerful ideas in a pictorial way. Our ray model of light, representing those ideas, has been able to deal with all sorts of situations, including quite complicated shadows caused by extended sources!

Generality, simplicity and power of ideas

Ideas in science ideally should have as much generality, simplicity and power as possible.

In this we have been quite successful so far, for light!

Science and math

Note that shadows themselves are an interesting phenomenon, but form a relatively small part of geometrical optics. However, the math techniques that apply to shadow behavior are general and powerful, in that they apply to many other topics in Light (and in fact to many other areas of science). Thus this section on shadows had a second agenda – to become proficient at some important "math for physics"! Both geometry and algebra. The topic of shadows was an ideal context in which to develop and apply these mathematical ideas. Thus this shadows section of the course is fairly long, because it is about math as well as shadows!

Your own reflections and insights

PROBLEMS

Shadows and the geometry and mathematics involved

The problems are grouped in types, as follows.

- Conceptual, phenomenon
- Qualitative conceptual
- Construction and simulation method. Qualitative and quantitative
- Ratio-logic method
- Math methods: quantitative and dependencies
- Combo type problems – combining various types of questions. Qualitative shadow behavior

---

QUALITATIVE SHADOW BEHAVIOR

1. How does shadow size vary as distances are adjusted?

A blocker card is placed between a light source and a screen, casting a shadow.

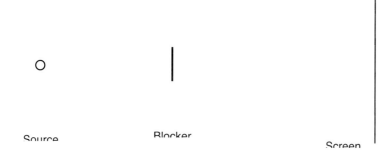

Source   Blocker   Screen

With the aid of our light model and diagrams, predict what will happen to the size of the shadow if:
- The screen is moved further away.
- The card is moved closer to the screen.
- The source is moved closer to the card.

2. How to increase shadow size?

A blocker is placed between a light source and a screen.

In what different ways can we *increase* the shadow size on the screen? By moving the screen, the card or the source? For each of these, say *which way* it must be moved. Support your answers with explanatory sketches.

3. Minimum shadow size

A blocker card is placed between a point source and a screen. Card and screen are parallel. The distances between these three items can be changed. What is the minimum size of shadow we can get, and what setup is needed for this? Is it possible to get a shadow smaller than the card? Why or why not? Illustrate your answers with sketches. (Note how the answers come out of our model of light behavior).

SHADOWS BY SIMULATION

4. What size blocker?

A screen is 90 cm from a point source, and we will place a square blocker card 60 cm from the source. What size card is needed to get a shadow 30 cm square? Solve by simulation.

5. To get a specified shadow

You have a point light source, a blocking card and a screen. Your aim is to get a shadow exactly 1.5 times the height of the card. Three cases follow. Do each by scale construction of a ray diagram.

a) If the source and card are at fixed positions, where must the *screen* be placed?

b) If the card and the screen are at given fixed positions, where must the source be placed?

c) If the source and the screen are fixed, where must the card be placed?

Instead of solving by constructing diagrams, we could solve the problem physically by simulating light rays using strings. That is, by making a physical model of the system. Try it.

Note

Drawing straight-line light rays on paper or using strings, are ways of *simulating* the actual behavior of the real system. That is, your construction acts much like light does! This is a powerful conceptual method and can be either qualitative or quantitative.

SHADOW PROBLEMS BY RATIO LOGIC

6. Getting height ratio from distance ratio

A card is between a point source and a screen, so that a shadow of the card is cast on the screen.

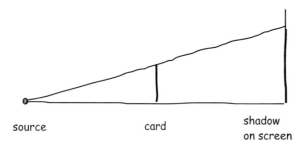

source   card   shadow
on screen

Suppose the screen is *1.5 times* as far as the card from the source. How many times taller will the shadow be than the card? Or putting it another way, what will be the ratio of the shadow height to the card height? Do this problem by ratio-logic as follows:

From the source, the screen distance is . . . . . times the card distance.
Therefore, the shadow height on the screen will be . . . . . times the card height.
Alternative wording using the term 'ratio' would be:
The ratio of screen distance to card distance is . . . . . .
Therefore the ratio of shadow height to card height will be . . . . . .

7. Shadow height using ratio logic method

A card of height 8 cm is 20 cm from a point source and a screen is 50 cm from the source.

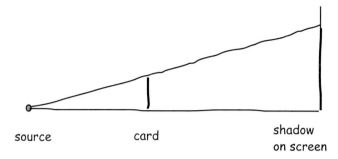

source   card   shadow on screen

What will be the height of the card's shadow on the screen? Do this problem by ratio logic, as follows:
The screen distance is . . . . . .cm and the card distance is . . . . . cm.
Therefore the ratio of screen distance to card distance is . . . .
Therefore the ratio of shadow height to card height will also be . . . . . .
The height of the card is . . . . cm, so the height of the shadow will be . . . . times this, i.e. . . . . .cm.

SHADOWS: MATH & CALCULATION

8. Calculate shadow height (1)

We take a problem already solved by the verbal ratio logic method, and now solve it using math. A card of height 8 cm is 20 cm from a point source, and a screen is 50 cm from the source.

What will be the height of the shadow? Do this problem *mathematically* as follows: Draw a rough sketch (not to scale). Label quantities on the sketch with symbols, known values and units. Obtain an algebraic relationship between quantities, from the geometry of the situation. Solve algebraically to get an expression for the shadow height. Then substitute known values and calculate a numerical answer for shadow height.

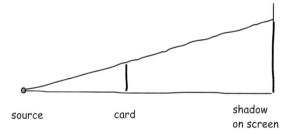

source   card   shadow on screen

9. Calculate shadow height (2)

You have a point light source, a blocking card and a screen. The blocker is H_b = 1 ft high and is a distance D_b = 2 ft from the source, while the screen is a distance D_s = 7 ft from the source. What will be the height H_s of the shadow on the screen? Solve this problem *mathematically*, using similar triangles and ratio relationships.

Note 1: Draw a *rough* sketch only, to show the approximate geometry. (A feature of the math method is that we do *not* have to draw accurate scale diagrams)!

Note 2: Work *algebraically* with symbols as long as possible, and only substitute numerical values at the end.

10. Variation on a theme –screen distance

We take exactly the same physical situation as in the problem above, but make a variation in the problem. This time the required shadow size is specified, and we ask what the screen distance must be to get this.

Formulate the modified problem *yourself* now, and then solve it mathematically, again starting with a labeled sketch.

11. More variations on a theme

For the same physical shadow situation, we can clearly set more problem variants, depending on which quantities we specify and which we want to solve for. Make up one or two variants for yourself, and solve them.

12. Overhead projector shadow

During a lecture presentation, the instructor steps in front of the overhead projector, so that her head is in the light beam. The shadow of her head on the screen is *five times* its actual size. If she is 3 feet from the projector, how far is she *from the screen*? Solve this problem *mathematically*. Use a rough labeled sketch to model the problem, show steps & logic, work algebraically, and only substitute numbers toward the end.

13. Height of high ceiling light in a hall

Here is a way to determine the height of a high ceiling light in a hall or theater, without having to climb up! Suppose you wish to know the height of a ceiling light in a high hall. Below the light a table casts a shadow on the floor. The table is 0.80 meters high. You find that the shadow of the tabletop on the floor is 1.2 times longer than the tabletop itself.
Draw a rough sketch of the situation.
Use the math of similar triangles to work out how high the ceiling light must be.
(Physics and math do have their practical uses! Also it's more satisfying to use one's intellect than trying to climb ladders…).

14. Estimating the size of the moon by holding up a finger

Here is a way to estimate the size of the moon. Hold your finger up at arms length, so it is just below the moon. Estimate how the apparent diameter of the moon compares to the width of your fingertip. (Of course one is much further away than the other, and we will use this fact).

Now sketch a diagram of the situation, not to scale. (Why not to scale?) Measure or estimate the width of your fingertip, and the distance of the finger from your eye. Now we know (by other means) that the moon is …. km from the earth. (Look this up). Using all these pieces of information, work out the diameter of the moon.

Can the method of construction to exact scale work for this situation? Explain. Note how the math method comes into its own for this situation… that's math power!

MATH DEPENDENCY QUESTIONS (qualitative and quantitative)

Note: The first few questions below are all variations on a theme, viz. various dependencies in shadow problems, related to the form of the algebraic relation between distances and sizes for shadows due to a point source.

15. Dependency of shadow size on screen distance (qualitative)

Consider the following (qualitative) question, expressed in terms of dependency:

An object is between a point light source and a screen. What will happen to the shadow size if you move the screen *closer*? Will it increase, decrease or remain the same? Or to put it another way, how does shadow size *depend* on screen distance?

Working from a diagram of the situation, set up the algebraic expression relating distances and sizes. Then answer the question by considering the *form* of the equation, in terms of how shadow size and screen distance are related. That is, if screen distance is decreased in the equation, what must happen to shadow size for the equation to remain satisfied?

16. Dependency of shadow size on screen distance (quantitative)

The previous question was qualitative: we simply talked of quantities *increasing* or *decreasing*. But the algebraic relation tells us more than that: it tells *exactly* how one quantity depends on another, i.e. it gives quantitative dependencies too. In our example, if the screen distance is reduced by a factor of 1.2, by what factor will the shadow height change?

DEPENDENCY QUESTIONS (MULTI-METHOD): USING BOTH PHYSICAL AND ALGEBRAIC REASONING

Shadow size depends on various distances – distance of source, blocker and screen. We are interested in knowing how shadow size varies with *each* of these. (E.g. if we increase the screen distance, how does shadow size change?).

We can see how one thing depends on another in three ways: i. by actually doing the real thing, ii. by modeling the situation and thinking physically about its behavior, and iii. by inspecting the form of the algebraic relationship between the quantities.

We can't always set up the real thing whenever we face a question (nor would we want to go to that trouble). So, we model the situation and think – either physically or mathematically. Aim to be good at both physical and algebraic reasoning!

Here are some practice problems to do both ways. You will recognize them as problems you did earlier one way only – now do them two ways.

17. How shadow size varies as distances are adjusted – by physical and algebraic reasoning

A blocker card is placed between a light source and a screen, casting a shadow.

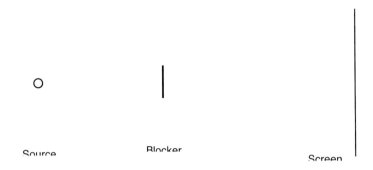

Source   Blocker   Screen

a) Draw suitable light rays to show how the shadow arises physically, and then use the geometry of the situation to set up an algebraic relationship between the following variables: D_b, D_s, H_b and H_s (as defined in the notes).

b) We want to predict what will happen to the size of the shadow if , , ,
    - The screen is moved further away.
    - The card is moved closer to the screen.
    - The source is moved closer to the card.

Do *each* of these in two ways, viz. i. physical reasoning, and ii. algebraic-dependency reasoning. Do you get the same result each way?

18. How can we increase shadow size? – answer by physical and algebraic reasoning

A blocker is placed between a light source and a screen.

In what different ways can we *increase* the shadow size on the screen? By moving the screen, the card or the source? For each of these, say *which way* it must be moved. Support your answers with sketches.

Treat each of these factors in two ways: i. by physical reasoning about the model, and ii. by looking at algebraic dependencies between variables.

19. Minimum size of shadow – by physical and algebraic reasoning

A blocker card is placed between a point source and a screen. Card and screen are parallel. The distances between these three items can be changed. What is the *minimum* size of shadow we can get, and what setup is needed for this? Is it possible to get a shadow smaller than the card? Why or why not?

Answer these questions two ways: i. by physical reasoning about a geometrical ray model of the situation, and ii, by looking at the dependency behavior of the algebraic relationship between variables.

Illustrate your responses with sketches. (Note how both answers ultimately come out of our model of light behavior).

SHADOWS COMBO PROBLEMS
(problems solved by a number of alternative methods)

20. Shadow calculation and shadow size adjustments

[Concept+Math+Qual+dependencies+real].

A teacher places a **circular** card between a small light source and a screen as shown.

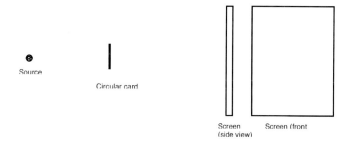

Source

Circular card

Screen
(side view)

Screen (front

a) Show on a *side* view how this setup produces areas of both light and shadow on the screen. Use many rays. Also draw what the screen looks like in *front* view, labeling screen areas as bright or dark.

b) The teacher now makes the card-to-screen distance exactly **twice** the source-to-card distance. If the card diameter is 10 cm, **calculate** what the shadow diameter will be. Refer your working to a labeled sketch. Afterwards, try it in practice: does your calculation agree with nature?

c) The teacher wants to get a **larger** shadow on the screen, so that the whole class can see it easily. She can move either the source, the card or the screen. State how each should be moved to get a larger shadow.

   Move the source to the… [ ] right   [ ] left

   Move the card to the…   [ ] right   [ ] left

   Move the screen to the… [ ] right   [ ] left.
   Then try it out for real: do things behave as you predicted above?

d) Set up the algebraic ratio equation for this shadow situation, using your diagram of labeled similar triangles. Confirm that the three dependencies in part (c) are also reflected in the algebraic equation. That is, show from the *algebra* what changes will lead to a larger shadow.

21. Where to put the screen? Done multiple ways

You have a point light source, a blocking card and a screen. The card is 2 ft high and is 4 ft from the source.

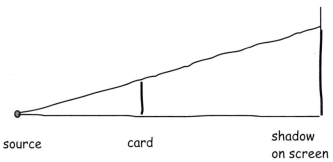

source   card   shadow
                                          on screen

Where must the screen be placed to get a shadow of height 6 ft?

a) Do this problem by scale construction, drawing light rays.

b) Do this problem mathematically, using triangle geometry and ratios. Use symbols first, then substitute values at the end.

c) *Dependency of* shadow *size on screen distance* If you wanted to get a *taller* shadow, would you have to move the screen closer or further back? Predict this both physically and mathematically, as follows:

    i. Predict physically using a diagram

    ii. Predict mathematically using your algebraic expression for required screen distance.

22. Shadow size by construction, ratio thinking and math

A small light bulb is 4 meters from a wall. You hold a 2-foot high pole upright, 3 meters from the wall.

a) What will be the height of the pole's shadow on the wall, in feet? Do this problem in three ways:

    i. by scale drawing and ray construction (simulating light)

    ii. by a logical sequence of statements involving ratio thinking

    iii. mathematically using triangle geometry, algebra and ratio calculations.

This problem gives some distances in meters and others in feet. Did you convert everything to the same units, or did you keep heights in feet and distances in meters? What are the characteristics of this type of problem that make it possible to work out answers without conversion? Support your answers by referring to some of the ratio calculations you made, with units explicit for each quantity.

b) What particular 'powerful physics ideas' about light and 'powerful math ideas' about geometry are relevant in this problem?

c) Young children might find the mathematical ratio approach too abstract and difficult. Do you think that the ray construction (simulation) method would allow young children to solve **all** of the quantitative shadow size problems that older students might solve using mathematics? Why? Or do you think some problems could only be solved mathematically?

Dependency shown by the equation

Explain how you can see *from the algebraic equation* that the shadow will get bigger if you move the screen further back.

23. Shadow problem tackled in multiple ways

A card 5 units tall is placed 10 units of distance from a point source of light. Where must a screen be placed so that the card's shadow on the screen is exactly 12.5 units tall? Tackle this problem in multiple ways:

    i. Do the real thing, with real light.

    ii. Simulate, using strings for light rays

    iii. Simulate geometrically by constructing a scale diagram on paper and measuring.

    iv. By ratio thinking using a verbal 'ratio logic' sequence of statements.

    v. By formal mathematics, using algebra and ratios. Annotate all steps.

Do these methods all give the same answer for the screen position?

What light properties and math ideas are implicitly involved for your solutions to work? That is, which *powerful physics ideas* about light were involved in your diagrams and calculations, and which *powerful mathematical ideas* were involved in working out the solution?

Reflect on the methods above, noting how they differ from each other and how they relate to each other. Become comfortable with each method. Note that some methods can be used with students of almost any age, others only with older students with math experience, but the same problem can be set for all.

24. How far away must the source be?

[This one is easy by construction but relatively hard by algebra & calculation]

An object OO' is 2 ft tall and is located 4 ft from a wall as shown. The object is illuminated by a point light source, and casts a shadow SS' on the wall. The shadow is found to be 3 ft tall, as shown in the scale diagram below.

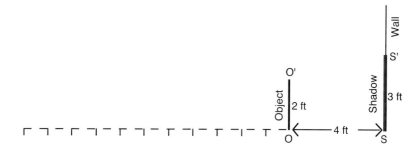

What must be the location of a point light source to produce a shadow this size? Determine this in two ways:

a) By ray construction on the diagram above. The dashed line is marked with one-foot intervals.

b) Mathematically using triangle geometry and ratio calculation. Use symbols first, then substitute values at the end. [Note: this is much harder algebraically than the other questions. Omit it if you like. Included for completeness].

c) Do your two answers agree? Which method do you think is the easier for this particular example?

d) If you wanted to get a *taller* shadow, would you have to move the source further away or closer? Predict physically, using a diagram.

GROUND-SHADOW PROBLEMS

Note:

So far, we have looked at shadow situations where the blocker and screen were parallel to each other. There is a different type of shadow situation, where the source, like the sun, is above the ground at some angle, and a vertical object like a tree or a person casts a shadow on the ground. The geometry is slightly different for this situation, but the physics concepts and math methods apply just as before!

25. Height of flagpole problems - create some like the example problem

Given some sample data (i.e., height of stick, length of stick's shadow and length of flagpole shadow), find the height of a flagpole.

26. Two students using shadows to find the height of a tree

In a class activity on shadows, the teacher sets the students the task of finding the height of a tall tree, using shadows. (No climbing allowed!).

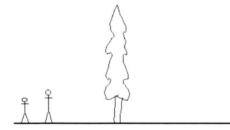

a) The class measures the tree's shadow on the ground to be 27 feet long. One student, Mary, who is 5 feet tall, finds that her own shadow is 3 feet long. Use math to show how she can calculate the height of the tree.

b) A smaller student Meg asks: I'm only 4 feet tall, not 5 feet like Mary, so if we had used me instead of her, wouldn't we get a different result for the height of the tree? How would you reply?

c) An hour later, the class notices that the tree shadow has gotten longer than the 27 feet they originally measured. What has happened to cause this? Illustrate with a diagram.

d) Student Joe asks: Now that the tree shadow is longer, if Mary did the activity and calculation again, wouldn't she get a different result for the height of the tree? How would you reply? As a teacher, reply with a series of questions to lead Joe to the answer, without just telling him.

e) Looking at his own shadow produced by the sun, Joe notes: 'the shadow edges are fuzzy, not nearly as sharp as the shadows we made in class using the maglite bulb'. Suggest why this might be.

PROBLEM CREATION

Make up shadows problems for yourself, of various types.

The instructor will give you feedback and may even use the most interesting questions in quizzes or exams or future courses!

Chapter 6

APERTURE EFFECTS

— another consequence of straight-line light travel

In an earlier section (shadows) we put an object in the path of light rays, the object blocked some rays and this gave rise to shadows. The effect was a consequence of the fact that light travels in straight lines. We now look at another consequence of straight-line travel, and the allowing or blocking of rays, namely the effect that occurs when light from an extended source goes through a small aperture.

We will use a large card to mask off most of the light coming from the source, except for a small aperture that allows some of the light to go through. This will give interesting and perhaps unexpected effects for extended sources! We will explore what happens and explain it using our ray model of light.

Further, it turns out that the aperture effect has rather important real-life applications to optical instruments, including for example the camera and the eye.

As usual, we will start our investigations in simple cases first, before looking at more complex situations or practical devices.

SECTIONS

PROBLEMS

6-1   EXPLORING THE EFFECTS OF AN APERTURE IN A MASKING CARD

Situation and focus questions

What will be the effect if we block off most light from a source but let some go through a small aperture? In particular, what will happen for an *extended* source of light?

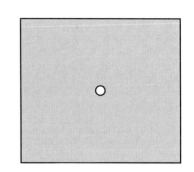

We will try it in a darkened room, first with a point source and then with an extended source. The aperture is a relatively small one, made by piercing a card with a pin. Small apertures like this are usually called 'pinhole' apertures.

(Note: for a large class demonstration one can use a larger aperture, as long is it is still small compared to the extended source used)

Point source and an aperture: what do we get?

Setup. First, we will use a point source (maglite). Set up a masking card with a pinhole aperture and place a screen beyond. Don't turn on the maglite yet.

Wondering: what do you think might happen? What might be seen on the screen? Ideas?

Try it. Now try it – what do you get? As expected or a surprise?

Well, that may have been expected and thus boring! So how about trying an *extended* source next?

Extended source and an aperture: what do we get?

Now we explore aperture effects for an *extended* source.

Setup. Get a light bulb with a "Λ" shaped filament to use as the extended source of light. (We want an extended source with a distinct shape). Don't turn it on yet, but set up the masking card with pinhole aperture as before, in a darkened room.

Wondering: What do you expect to see on the screen when you turn on the Λ-shaped source? Ideas?

List the various class predictions:

Try it

Now try it – what do you get?

Was it expected? Or a surprise?

Large scale demonstration

The lab also has a large scale class demonstration of the effect. The *large "Λ"* shaped source is make of two straight filament bulbs at an angle, in a box with one side open. The aperture is about 1 cm diameter in a piece of board. This produces a nice large picture on a wall or screen in a darkened room, visible to the whole class.

Puzzle to explain

So what's going on? Can you it out and explain the effect of an aperture on light from an extended source? Illustrate with a ray diagram.

Ray diagram to explain the effect of an aperture on light from an extended source.

Functions of the mask and the aperture

So, what are the *functions* of the mask and the aperture in this situation? Illustrate on your diagram what each of them is doing.

Would you agree that the mask blocks off most of the light from the extended source, while the aperture acts as a sort of 'ray selector', allowing just some of the rays from each point of the source to get to the screen, and thus 'controlling' the emitted light, to get a useful result?

Note: The mask is as important as the aperture in its action on light from the source. To see this, imagine what would happen if you removed the mask, so that light from the extended source could go 'unchanneled' to the screen? Predict and then try it.

What we want is a system that will block some light rays and allow others, and a mask with an aperture is just such a 'ray selector' device! Simple but effective.

String models

You can make a string model of how apertures operate with light. Use different colored strings for light rays. Tape one end of each string to a point on a source object, pass the string through a hole in a card, then tape again where it meets the screen.

The lab has a large scale string model for you to look at also!

Anticipatory note on applied geometrical optics and optical instruments

Light selection and manipulation as a basis for optical instruments

This may be a good time to look ahead and say more generally that applied geometrical optics is mostly about selecting and manipulating light! Various basic systems such apertures, mirrors, and lenses, can function to *select* some of the light rays from a source and *manipulate* them in a controlled way (. e.g. by changing their directions). This causes useful optical effects, in particular it can produce an **image** of a light source. Such images, obtained by light transformation, will generally be a different size from the original source, and occur in a different location. This turns out to be immensely useful in optical instruments such as the camera, projector, telescope, microscope, etc. When using such devices, you are looking not at the original object, but at an *image* of the object obtained by light manipulation.

Thus far we have encountered our first and simplest light manipulation device, viz. blocking mask with aperture. We found that it can produce pictures on a screen, and that the picture size can be controlled. Other optical devices are more sophisticated, but they all involve light ray control in one way or another.

Reflections on modes of thinking

Initial intuitive thinking

How were you thinking when you made your first prediction for the Λ–filament and aperture?

And how are you thinking now?

If you had been thinking from basics and principles right from the start, using the known properties of light behavior, could you perhaps have predicted correctly at the start?

Scientific (principled) thinking

Note that even before learning anything about apertures, you already had all the basic pieces of science knowledge necessary to tackle the situation! The context may have been new and unfamiliar to you, but as long as you *worked from basics* and *applied principles and models*, you could have worked out what was going on and what effects should occur. This shows the importance of a scientific approach in your thinking.

It also highlights the power of our 'powerful ideas' and 'powerful model' foe light. These basic ideas and model can predict or explain what happens in **new** situations, which we didn't even have in mind when we initially developed the ideas and model.

Discussion notes

6-2 APERTURE EFFECTS: EXPLORING QUALITATIVE BEHAVIOR

Now that we understand the *mechanism* of the aperture effect, we should be able to *explain* and *predict* the type of behavior that we get when we make *changes* to the setup. That is, as we make changes to various distances, what happens and why?

First, sketch the basic aperture setup, with extended source, mask with aperture, and screen.

General aperture setup

What is the effect of source orientation?

We now have an idea of how apertures work to produce pictures. If this idea is correct, what should happen if you change the *orientation* of the source? *Predict* what you would see on the screen if you turn the source upside down, or rotate it 90°. →

Reasons? →

Then try it out and report.

What will happen if you adjust various distances?

If your ideas about apertures are correct, you should be able to predict what will happen to the reproduction if you vary the setup, e.g. if you change some of the distances between source, aperture and screen.

Prediction problem: – what if we move the screen further away?

a) Think for a moment: if your proposed mechanism is correct, predict what should happen to the reproduction if you *move the screen further away*.

Give your reasoning illustrated with a diagram.

Then try it out in practice with a real setup, and report:

b) Predict whether you should be able to get pictures of *any* size, i.e. pictures which are larger, smaller or the same size as the extended source.

Explain and illustrate

Then try it out in practice (to check that you and nature see eye to eye, so to speak!).

Report:

Note

Note that we are tackling these questions by using lines on diagrams to represent light and to *simulate* how it behaves. Simulation is dealt with in some detail later.

Homework and home experiment

A pinhole aperture is easy to make at home. Usee a pin to make a small hole in a piece of aluminum foil; the hole has cleaner edges with foil than card. A clear light bulb with a visible filament of some shape can serve as an extended source. Using these, try out aperture effects yourself at home and explore the behavior.

☆
Homework – do on separate page

Here are some predictions and home experiment tasks to work on.

a) Predict the effect of moving the *source* further back. Illustrate and explain. Then test it as a home experiment.

b) Predict the effect of moving the *aperture* further from the source. Illustrate and explain. Then test it as a home experiment.

Report on these.

6-3 REFLECTING ON MODELS AND MECHANISMS

At this stage, we have two potent tools in our quest to understand aperture effects – a *model* of light behavior, and a *mechanism* for how a mask and aperture acts on light to give a reproduction of an object. Armed with this model and mechanism, we can confidently tackle virtually any new situation about apertures! We can explain effects, make predictions and solve problems.

Developing *models and mechanisms* for phenomena is a large part of what science is about. As to your own understanding, if you learn to think in terms of models and mechanisms, you will be able to figure out situations for yourself from basics – and get things right. You will be able to think through any problem, familiar or unfamiliar, instead of hoping to memorize lots of particular results – and often getting things wrong. In teaching too, it is best to emphasize underlying principles, models and mechanisms, rather than lots of disconnected facts, formulas and results. The former will lead the learner to real understanding of the subject matter and an appreciation of the nature of science.

Since this way of understanding nature is so important, it is best to take the trouble to elaborate our light model and aperture mechanism, in both words and sketches.

Model – of light being emitted from an extended source	Mechanism – of mask and aperture acting on the light from the extended source

NOTES

6-4 APERTURE EFFECTS BY SIMULATION
– using ray diagrams and string models

Models, mechanisms and simulations

Now that we have a model and mechanism to explain how aperture effects arise, we can *simulate* the situation. We can simulate light diagrammatically by rays drawn on paper, or simulate it physically by strings.

To solve aperture problems or make predictions, we can use the *simulation,* anytime and anywhere – we don't actually need to have a real source and aperture available. Of course we assume the simulation will behave like the real thing!

Simulations can be used in both qualitative and quantitative problems.

The best way to illustrate the simulation approach is to tackle some specific problems.

Example problems: to be solved by simulation

You can use scale diagram constructions, or string simulations, or computer simulations, to represent light ray behavior, and hence solve aperture problems (both qualitative and quantitative).

1. Qualitative problem – what happens if the screen is further away?

A mask with aperture near an extended source of light produces a picture of the source on the screen. Make a ray sketch showing how this works, then use it to explain what will happen if the screen is moved further away. Note that here you are using lines on paper to represent and simulate light behavior.

You could do the same thing using a model with strings as rays.

After answering, try the effect for real, using a filament bulb, mask, aperture and screen – check how the picture does behave when you move the screen back, to confirm what you predicted by simulation.

2. Quantitative problem – where should the aperture be?

An extended source is 90 cm from a screen. Make a ray construction to find out where a mask with aperture should be placed to get a picture which is exactly **twice** the size of the extended source.

Note

Simulation is clearly a powerful method for solving problems, and conceptually valuable since it represents the light behavior directly.

But be aware that in some cases the comparative distances involved may preclude scale constructions, in which case we can use math calculations instead, as dealt with in the next section.

6-5 APERTURE EFFECTS BY MATHEMATICS

We now know how aperture effects arise, with some of the light rays from an extended source being 'selected' by a mask and aperture arrangement. We also understand the qualitative behavior of this setup, i.e. how and why the picture size changes as distances are varied. Now it's time to turn to quantitative math treatments. This will be a particularly powerful tool in solving problems.

Transfer challenge

Can you **transfer** your existing math knowledge to the aperture situation?

It turns out that you already have all the math ideas needed to tackle aperture problems mathematically! These ideas are in the 'math toolkit' we developed for shadow problems. There, the powerful tools were the math of similar triangles and side ratios. The challenge will be to transfer that knowledge to a new situation! Let's try.

Typical geometry for apertures

Take a typical apertures situation, with extended source, mask with aperture, and screen.

Draw a ray diagram

You'll see that similar triangles virtually jump out at you! (Justify why they are similar),

The geometry is different from that for shadows. But we still have two similar triangles, so we can go ahead and work with side ratios!

> Ray geometry for aperture effect

Side ratio equation

With these few hints, its over to you.

Label the relevant quantities on the triangles.

Develop an algebraic ratio equation relating distances and sizes for this situation.

> Ratio equation

Math power!

Now, armed with both geometry and algebra (diagram and equation) you should be able to solve any apertures problem! That is, given three of the four quantities involved, you should be able to find the fourth. Notice how you do not have to learn any new math – you can bring it over from the shadow section, and use it for apertures too!

EXAMPLE PROBLEMS – to solve using math

Overview

Aperture problems can be solved mathematically using the geometry of similar triangles and the algebra of side ratios. You can set up an algebraic relationship between object size, picture size, object distance and picture distance.

Given any three of these quantities, you can solve for the fourth mathematically.

Aperture problems using math

Take some of the aperture problems solved earlier using simulation (accurate ray construction) and now do them using math instead.

The instructor will advise and guide. For each problem chosen, check if the answers you get using simulation and math agree.

T

6-6 APERTURE EFFECTS: DEPENDENCIES
– by 'reading' algebraic equations

The algebraic relation for aperture effects, besides being useful for solving numerical problems, also tells us how one quantity **depends** on the others.

Just as we did for shadows, we can 'read' the algebraic equation, to see how one quantity will change if we vary one of the others.

Example questions on algebraic dependencies

We can take examples from the earlier section on qualitative behavior, which we answered at the time by simulation (ray diagrams to scale). Now we will work instead from the algebraic equations, and answer the behavior/dependency question by 'reading' the equation.

At the end we can compare the answers we get from the two methods.

Here are the questions again, from the earlier section. For a basic aperture setup,

- a) What will happen to the picture if you move the screen further back?

- b) What will happen if you move the object further back?

- c) What will happen if you move the aperture closer to the screen? (This is more complicated; why?)

Compare your answers to those obtained earlier in section 6.4 on qualitative behavior, where we used ray diagrams to simulate behavior

You can also check your answers against nature by trying it for real.

Notes & reflections

Note how the algebraic relationship for aperture effects was useful in two ways: i. for numerical solution of a particular problem, and ii. for representing dependencies in the system.

6-7 WHAT ARE THE EFFECTS OF APERTURE SIZE?

– Developing further knowledge

So far we have only used **small** (pinhole) apertures. The question arises: what would be the effects of a larger aperture? What might change on the screen, and why?

Think, and give some ideas and reasons:

```
Initial ideas



```

Now try it in practice. Use a card with two different-sized apertures, and compare the pictures produced by each. Record your results.

Do the results make sense? How can we explain what happens? We surely should be able to use our model of light behavior to account for this!

For an extended source, and a tiny pinhole aperture, draw a diagram showing how a picture is formed on a screen. Include light coming from several points on the source.

In a second diagram, use a somewhat larger aperture. Now, a somewhat larger *cone* of rays will go through the aperture from a point of the source to strike the screen. What effect will this have on how that particular point is pictured on the screen? What effect will the larger aperture size have on the appearance of the picture? [Instructor can supply diagram from Wheeler and Kirkpatrick to illustrate this effect].

Increasing the aperture size also has a *second* effect on the picture. Can you suggest what it might be? Did you observe it in practice?

Suggest *advantages and disadvantages* of using small or large apertures. (We will see later how cameras manage to use large apertures and still get sharp pictures!)

6-8 KNOWLEDGE SCHEMATA FOR APERTURE EFFECTS

Think back on what you have learned about the aperture effect, and identify the essence. Then devise useful knowledge subassemblies for the topic.

You have a head start since we have been using a ray diagram to show 'model and mechanism' of how a picture of an extended source is produced. This can already serve as a subassembly. Would you agree that it combines the purposes of phenomenon diagram, and principles & procedures diagram? Can it also serve to deduce behavior when various factors are imagined varied? If so it can serve as a behavior (variation) diagram as well.

Draw an annotated subassembly diagram to serve these purposes:

This diagram can serve as your mental model when thinking about the effect or solving problems.

Do you want a separate subassembly for effect of aperture size?

6-9 PRACTICAL APPLICATIONS OF APERTURES:
Various optical instruments which make use of the effect

Introduction

Does the aperture effect have real-life uses? Let's imagine being an inventor!

First of all, lets look at a demonstration of the basic effect again.

[Demonstration withextended source (say a light bulb filament,) and aperture, getting a picture of the source on a screen].

So, can you think of practical applications for this? Can you use the effect to *invent* something? See where your imagination will take you before reading further. Share ideas.

Ideas:

Devices

It turns out that there are various optical devices based on the principles of straight-line light travel and ray-selection by an aperture.

We will discuss the following devices: the slide projector, the photographic enlarger, the 'camera obscura', the pinhole camera, the regular camera, and the eye, etc. These all make use of the aperture effect in producing *reproductions* of extended sources, though they also have more sophisticated aspects, due to lenses. The reproductions may be produced on a screen, or on photographic film, or on the retina, as the case may be.

Primary and secondary sources of light for these devices

Note that an extended source of light does not have to be an active (primary) emitter of light, it can also be a *secondary* emitter. That is, light illuminates the object, and some light is then emitted from its surface. Any illuminated object can thus serve as an extended light source! For example trees, people, buildings etc. can all serve as extended light sources. In that case, a mask and aperture can produce a corresponding picture of such objects on a screen. You can already imagine how useful this might turn out to be! We can produce pictures of any illuminated object or scene!

OPTICAL DEVICES BASED ON THE APERTURE EFFECT

There are a number of useful optical devices which make use of the aperture effect (usually in conjunction with lenses to obtain sharp focused images).

It is useful to group them under *projection* devices and *camera* devices, though the same principles apply to all and the distinction may be arguable.

A. PROJECTION DEVICES

1. The primitive projector

Just a mask and aperture on its own functions as a primitive projector! After all, given a bright source, this simple equipment produces a 'picture' of the source on a screen! The source has been 'projected' elsewhere. So, our demonstration of the aperture effect was also a demonstration of a primitive 'projector'. Seem like this could be refined to be useful?

2. The slide projector

Demonstration and diagram of manual slide projector. Parts.

Principle of operation: Aperture ray-selection. Role of lens.

Orientations of extended source and projected picture.

3. The overhead projector

4. The photographic enlarger

B. CAMERA-TYPE DEVICES

1. Camera obscura

Dark room, hole in window blind. (What does camera obscura mean?)

Observed effect. Orientation of image.

Explanation and ray diagram.

2. Pinhole camera

Box with pinhole. Shoebox or 'Pringles' chips box. Orientation of image.

Principle of operation: ray selection by aperture.

Practicality?

3. Regular camera

Diagram. Principle of operation. Ray-selection by aperture plus lens action

Inspect an open camera. Orientation of image.

Role of the lens – brief.

5. The eye

Diagram of eye. Labeled.

Principle of operation. Aperture ray-selection.

Orientation on retina. Mechanism of seeing: retina, optic nerve, brain.

Roles of cornea and lens (brief)

Note on operation of all these devices, and the use of 'model and mechanism'

Note that at heart, these devices all depend on 'selecting' and hence controlling the light emitted in all directions from every point on an extended source. The selector is a mask with aperture. The mask blocks off light rays that we don't want, while the aperture selects rays that we do want, to pass through on to the screen or film. Thus any given point on the screen will receive only light from a given place on the extend object, rather than receive all the light the all parts of the object.

To understand this further, imagine the mask and aperture is removed in each case; so there is just the source (extended object) and the screen. What would be seen on the screen?

6-10 REFLECTING BACK ON KNOWLEDGE

Note how shadows and apertures both use the same mathematical technique! Similar triangles and side ratios feature in both phenomena!

Draw typical shadow diagrams and aperture diagrams side by side. Note how triangles occur in both. Compare.

How are they the same? How are they different?

Shadow diagram   *Aperture diagram*

Comparison:

Notes

It is interesting how the same geometry and math ideas apply to both phenomena, and are used for problem solving in both. Thus although the situations are different, the underlying mathematical methods are common to both! This shows "math power"!

146

6-11 FORMULATING OBJECTIVES

Broad objectives are commonly stated at the beginning of a section. In our case however, with an inquiry approach to developing the topic by investigation, we identify a set of very detailed objectives and compile them here at the end of the section. These objectives should be useful and most meaningful when you review the section.

This section is certainly about the topic "light and apertures", but you will have seen it's really about much more than topic content! It's also about models, abstraction, representations, physical insight, qualitative reasoning, mathematics, calculation, etc - all involved in understanding this particular phenomenon. Thus it's about the topic, physics, math, and thinking – and how they connect. We learn how to *model* physical systems to reflect their essence, and how to *represent* system behavior in various ways, e.g. diagrammatically or mathematically. These powerful and general techniques serve us in many topic areas. We learn them in the context of this particular topic and use them to solve problems. Below is a list of very detailed objectives for the section, both content and beyond.

Objectives for learners:

a) Wonder and predict what might happen when light from an extended source encounters a masking card with a small aperture, and goes through to a screen beyond. Then try it and observe.

b) Explain how such a setup works to produce a 'picture' of the source, using our ray model of light emission and travel, and why the picture is 'inverted'.

c) Simulate/model how the effect arises, by string and by ray diagram.

d) Explore how picture size changes as distances change, and explain the behavior using qualitative physical reasoning.

e) Use scale construction and measurement to solve quantitative problems involving sizes and distances.

f) Set up an algebraic relationship between sizes and distances, based on similar triangle side ratios.

g) Solve quantitative size-distance problems using algebra and calculation.

h) Deduce qualitative dependencies between sizes and distances from the form of the equation.

i) Solve both qualitative problems and quantitative problems.

j) Do problems multiple ways, e.g. by ray diagram construction, verbal 'ratio-reasoning' and algebra.

k) Appreciate that the effect, and its behavior, follows from our model of light behavior, and that we could have predicted it by thinking in terms of model and mechanism.

l) Reflect on possible 'everyday' ideas and modes of thinking about the effect, and compare with 'principled' scientific thinking, i.e. thinking in terms of models and principles.

m) Appreciate that an extended source is required, and that mask and aperture together can be seen as acting as a 'ray selector', to bring about both a picture and inversion.

n) Recognize that problems that seem different or unfamiliar on the surface may involve common underlying principles, and be able to view problems in this way.

o) Explain how aperture size affects picture sharpness and brightness.

p) Show how various optical instruments (e.g. camera obscura, pinhole camera, regular camera, and the eye) rely on the mask-with-aperture mechanism to produce pictures of extended sources. And how they use lenses to get a sharper picture.

q) Solve problems on these optical instruments, recognizing that they are just aperture problems in a practical real life context.

r) *Cross-topic reflection:* Realize that although aperture effects for an extended source may seem different from shadow behavior from a point source, the underlying model and math is much the same! Both involve ray diagrams, similar triangles, size ratios, scale constructions, and algebra. Thus our earlier approach and analysis for shadows transfers right over to aperture effects!

s) Recognize ultimately that this section is about more than "facts on apertures". It's also about more general aspects of science, things like models, representations, reasoning, mathematics, dependencies, problem solving, etc. Become 'literate' at it all!

t) Add some objectives of your own…

PROBLEMS

Aperture effects

PROBLEM CATEGORIES

- A. THE BASIC PHENOMENON
- B. APERTURE PROBLEMS BY SIMULATION
- C. EFFECTS OF APERTURE SIZE
- D. APERTURE PROBLEMS BY MATH CALCULATION
- E. APERTURE INSTRUMENTS QUESTIONS
- F. COMBO QUESTIONS

A. THE BASIC PHENOMENON - what an aperture does

1. **Lead-in question: two small point sources and an aperture**

   The figure shows two small (point) sources of light, A brighter than B, in front of a board with a pinhole aperture, and a screen beyond. A front view of the screen is also given.

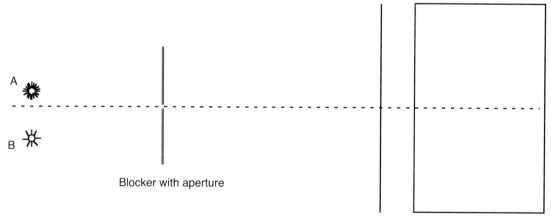

Blocker with aperture

Screen   Screen seen from front

a) Draw suitable light ray paths to determine accurately what would occur on the screen. Also draw what the screen would look like seen from the front. Say how the brightnesses on the screen would compare.

b) Source A is above source B. What is the case on the screen? State why this occurs.

c) If there were **more** point sources, what would you get on the screen? (You may recognize this as a lead-in to understanding an **extended** source and an aperture!)

d) If the blocker with pinhole were removed altogether, what would be seen on the screen? Explain. You can see what the aperture is doing for us then!

Comment: This question also highlights the way that an aperture in a masking card operates as a 'ray selector'! It 'selects' just a narrow beam of rays, of all the rays being emitted from a source, and allows this directional beam through while blocking all the other being emitted in other directions.

2. What will be seen on the screen?

A glowing lamp in the shape of an upright arrow is in front of a card with a small round aperture in it, with a screen beyond, as shown.

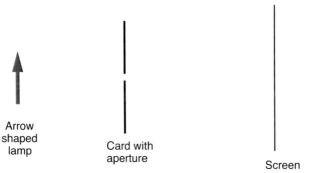

Arrow shaped lamp

Card with aperture

Screen

What will be seen on the screen?
   A.   A small round spot of light.
   B.   An upright arrow.
   C.   An upside down arrow.
   D.   Uniform illumination, with no arrow shape seen.   2

The card with aperture is now removed altogether. What will be seen on the screen?
   A.   A small round spot of light.
   B.   An upright arrow.
   C.   An upside down arrow.
   D.   Uniform illumination, with no arrow shape seen.   2

3. Ways to get a bigger picture with a pinhole aperture system

Suppose you are doing a pinhole aperture demonstration for your science class, showing a projection of an extended source, using the setup shown.

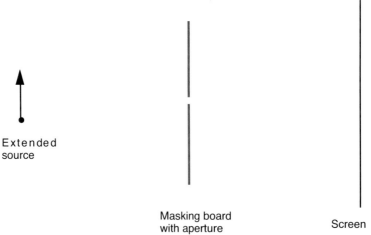

Extended source

Masking board with aperture

Screen

You decide you want to get a bigger picture on the screen, so that students at the back of the class can see it easily. The source, aperture and viewing screen are all movable. List the different adjustments you could make by moving things, to get a bigger picture on the screen. In each case give the direction of the adjustment needed.

Then explain how ONE of these possibilities works, with the aid of a suitable diagram.

B. APERTURE PROBLEMS BY SIMULATION
(Ray construction, strings or computer)

4. Extended source and *two* apertures: principles and practice

Suppose you have an extended source (e.g. V-shaped filament, arrow etc) and *two* pinhole apertures in a masking board as shown below.

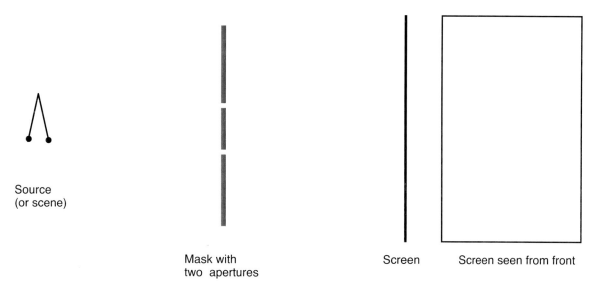

Source
(or scene)

Mask with   Screen   Screen seen from front
two apertures

What would be seen on the screen? You can approach this problem in either of two ways:

• *Principled prediction first, then experimental check.* Predict what would happen, using a model and principles, and then check by trying it out.

• *Experiment first, then explain using principles* Here you try it first, see what you get, and then explain using a model and principles.

In either case, both theory and experiment are required. Do it at home and report.

Discussion for teaching

Note that we have asked for both theory and experiment in this task. This is unusual in science courses; they usually limit problems to paper tasks. But we think it is important to link theory to actual phenomena and confirming tests.

If you were teaching this to your own students, which of the two approaches above would you prefer? Prediction from principles first, or experiment first? Why? Does one or the other, or both, represent kind of thing that happens in doing science? Could either teaching approach work well, if planned well?

5. To get a picture twice the size of the object

Using a pinhole aperture, we want to get a picture on the screen which is exactly **twice** as big as the actual object (source). Suppose the source (arrow) and screen are in fixed positions, as shown.

Extended
source

Screen

a) Where should we place a masking card and aperture, in order to get an arrow picture **twice** the size of the arrow? Do this problem by accurate ray construction on the diagram.

b) You can also solve this physically using strings. For this a convenient practical choice might be placing source and screen 60 cm apart.

c) You can also do the problem mathematically, using similar triangles and math ratios. Again, choose 60 cm as a convenient distance between source and screen, and calculate where the aperture must be located to get a picture twice the length of the source arrow.

Natures Test

Then try out the real thing. Set things up, and see what is required to get a magnification of two.

C. EFFECTS OF APERTURE SIZE

6. What difference does pinhole size make?

This question is about the effect of pinhole size. The figure shows a **point** source of light, two different sized holes in a board, and a screen beyond. Many light rays come out from the source.

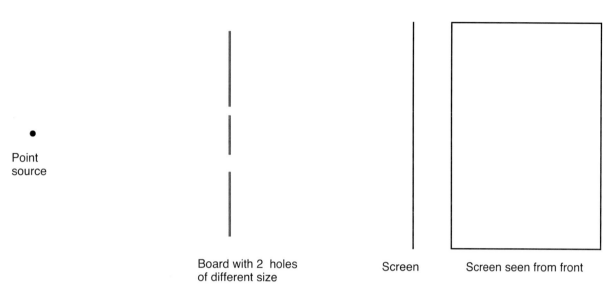

Point
source

Board with 2 holes
of different size

Screen

Screen seen from front

a) By drawing a number of rays, show how both *shadow* and *lit* areas are produced on the screen.

b) The screen is also shown viewed from the front. Draw what you will see on the screen, to match your side view.

c) You will see how a tiny point source gives a spot of some extent on the screen, not just a point. And clearly the spot size will depend on the aperture size. Let us carry this idea over to an extended source: do you see why a pinhole camera will give a reasonably sharp picture of a scene, but if you make the aperture bigger the picture will be brighter but fuzzier? Explain.

7. Seeing though a tiny gap between your fingers. Personal experiment.

In this question you first observe an effect for yourself and then explain it. The instructor will show you how to do it.

EITHER: If you are nearsighted and wear spectacles, take them off and look at writing on the board, which should appear too out of focus to read.

OR: If you have good sight, then bring the small print below closer to your eyes, until it is too fuzzy to read.

> Small print for observing the effect.
>
> Look at this small print from very close, through a tiny gap between your fingers

Next, arrange your fingers to make a tiny gap between them. Hold the gap close to one eye and look through it. You should find that you can see the letters and words much more clearly! In fact if you make the gap slightly larger or smaller you can immediately see the effect on the sharpness.

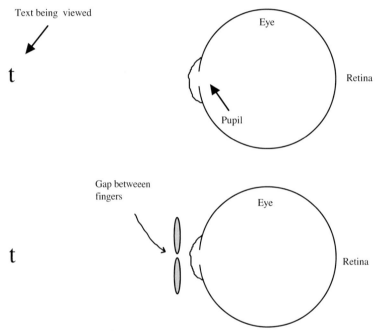

Explain why looking through a tiny hole makes the words look sharper. Illustrate your answer with suitable ray diagrams. The diagrams below may assist you. (Hints: i. The tiny gap is smaller than the pupil of your eye, and ii. Consider what happens to rays from a point on the letter 't' shown).

(*Note.* This is a real technique used by nearsighted children in classrooms, to read the chalkboard if they don't have glasses. Somehow they discover it for themselves. Two instructors of this course actually did this in their childhood!)

Note also that a person over 50, who can't read small newsprint easily, will often use a very bright lamp. This has nothing to do with needing more light, rather it causes the pupils to contract – explain why this causes the print to get clearer. Just consider a single point at the edge of one letter, and explain why it is seen fuzzier in dim light than in bright.

D. APERTURE PROBLEMS BY MATH CALCULATION

8. Instructor problems

Instructor will provide and discuss

9. Using a pinhole in a card to find out stuff about the sun

a) How does the diameter of the sun compare to its distance away?
You can find out, even though the sun is so far away, by doing a simple home experiment with a card and pinhole! The sun is an extended source, and so we can use a pinhole to get a reduced picture of it on a piece of paper, and work with that!

First, try the home experiment. Make a pinhole in a piece of thin card. Hence produce a picture of the sun. Caution: don't look at the sun itself, just its picture produced on the paper.

Use your knowledge of optics, measure what you need, and hence answer (a) the ratio of the sun's diameter to its distance away.

b) Then work out (b) the sun's actual diameter, assuming that we know by other means that the sun's distance away is 1.5 x 1011 m. i.e. 1.5 [11] m.
Answer 1.4 [9] m

Reflection:

This is an example of 'idea power' in science! Knowing about light and aperture effects, and using basic triangle geometry, we can find out things about the sun! While doing a simple activity in our own back yard! The ideas and principles are simple but powerful, applying to all kinds of situations.

E. APERTURE INSTRUMENTS QUESTIONS
(Camera obscura, pinhole camera, camera, eye)

10. Pinhole aperture camera

The diagram shows a pot plant and a basic pinhole camera (box with small aperture and film).

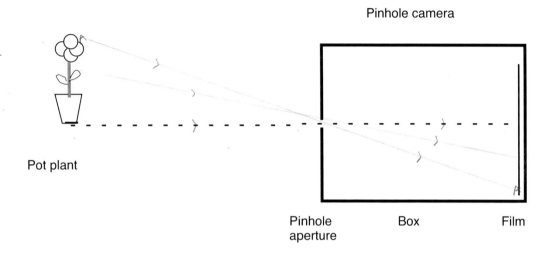

Pinhole camera

Pot plant

Pinhole aperture   Box   Film

a)   Draw at least three light rays to show the system works to get a picture of the pot plant on film.
                                                                                              (3)

b)   i.   Will the picture on film be the same way up as the object, or upside down?   (1)

      ii. Is this a drawback of pinhole cameras for getting photographs, or not? Explain.   (1)

      iii. Would a regular camera (with lens) have this same orientation of the image on the film, or
           not? Yes [ ]   No [ ]   (1)

c)   The pot plant does not give out light of its own accord. How then can one get a photograph of it, if
      you need light to expose the film? Explain how this all works.

d)   The pot plant is 20 cm high and is 60 cm from the pinhole. The camera box is 40 cm long.
      Calculate the height of the pot plant picture on the film. Do this as follows. Label the relevant
      distances with symbols and values on a sketch. Obtain an algebraic equation relating the
      quantities, and then proceed numerically to get the result.

e)   Find the **height** of the picture on film by careful construction and measurement on the diagram,
      assuming it is draw to scale. Indicate your method and results. (6)

f)   If the plant was moved further from the camera, the size of its picture on film would

           increase [ ]   decrease [ ]   remain the same [ ]   (1)

g)   If you used a **larger pinhole** in the camera, how would this affect the sharpness of the picture, the
      size of the picture, and the time required to get sufficient exposure on the film?

      Picture quality   Sharper [ ]   The same [ ]   Fuzzier [ ]   (1)

      Size of the picture: Smaller [ ]   The same [ ]   Bigger [ ]   (1)

      Time for exposure: Shorter [ ]   The same [ ]   Longer [ ]   (1)

h)   On what basic "Powerful Ideas" **about** light does the operation of a pinhole camera depend, to
      make pictures of objects? State them.   (x)

11. How do pinhole cameras work, and why are they not in common use?

How does a pinhole camera work, to take photographs of a scene? Explain, with the aid of a labeled diagram. On what principles or properties of light does its operation depend?

If pinhole cameras work, and are very cheap, and give quite a sharp image as long as the aperture is small, why are they not in common use? Why do we use expensive regular cameras with lenses instead?

A regular camera also has an aperture, much bigger than a pinhole, and a lens next to this aperture. Why does it use a relatively large aperture, and why does it use a lens?

12. Pringles pinhole camera

The figure shows a pinhole camera we tried out in the lab, made of a Pringles potato crisp box and a viewing screen which can slide in and out using a handle.

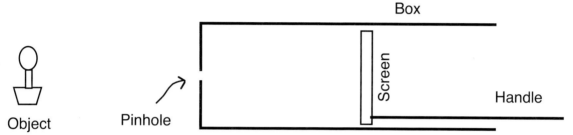

If the screen is pushed further into the box, i.e. closer to the pinhole, what happens to the size of the picture formed on it?

    a)   Stays the same
    b)   Gets bigger
    c)   Gets smaller

Explain by sketching on the diagram. (10)

13. The eye: size of the picture on the retina

Similar questions to pinhole camera questions!

E.g. object could be a tree 18 m tall and 120 m away, or a building 50 m tall and 1/2 km away, and give the distance between pupil (aperture) and retina, ask for size of picture on retina.

Variations, depending on what is known and what asked.

F. COMBO QUESTIONS - questions combining various aspects

14. Pinhole camera reproduction

A. Show with a diagram and light rays how a pinhole camera produces a reproduction of an object on film.

B. If the reproduction is formed upside down on the film, is this a problem for taking or viewing photographs? Explain.

C. A woman 5' tall is standing 18' from a pinhole camera. The camera box is 6' long; i.e. the film is 6" from the pinhole. Calculate, using triangle geometry, the height of the 'reproduction' of the person on the film.

D. If the woman walks further away, what if anything happens to the size of her reproduction on the film? Explain in two ways:

A. By using a ray diagram

B. By referring to dependencies in the algebraic equation you used in (c), noting how reproduction size varies with object distance.

15. Possible arrangements to get a specified picture size

You wish to demonstrate the pinhole aperture effect on a large scale to your class. You have a neon light 6" tall. Show the arrangement you would use to get a 2' high projection of the neon light on the screen. That is, decide where you will put the neon bulb, aperture and screen, giving the distances they are apart.

You should do this problem both by ray construction and by math calculation.

 Is there just one solution to this problem, or many? That is, is there more than one arrangement that will give the required picture size? Discuss. Explain how this possibility is reflected in the ray diagram approach and in the math equation approach.

APERTURES - MULTIPLE-CHOICE QUESTIONS

16. Getting a bigger picture

Suppose you are doing a pinhole aperture demonstration for your science class. You show a projection of an extended (arrow) source, using the setup shown.

You decide you want a bigger picture on the screen, so that all can see. The source, aperture board and screen are all movable. To get a bigger picture:

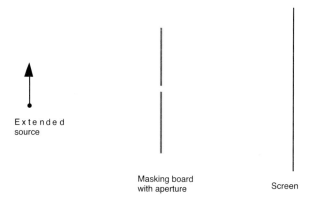

Extended
source

Masking board
with aperture

Screen

The source could be moved to the: [ ] right   [ ] left   (1)
The aperture board could be moved to the: [ ] right   [ ] left   (1)
The screen could be moved to the: [ ] right   [ ] left   (1)

17. Aperture effects – multiple MCQ

When light from an extended source passes through a pinhole aperture, a reproduction (picture) of the object is formed on a screen beyond.

Draw a ray diagram of the situation. Then mark the correct answers to the following questions.

a)	Compared to the object, the picture will be formed:	[ ] upside down [ ] right way up
b)	If the distance from object to aperture is greater than that from aperture to screen,	the picture will be [ ] larger than [ ] smaller than the object
c)	If the distance between aperture and screen is increased	the size of the picture will [ ] increase [ ] decrease [ ] remain the same.
d)	If the object is moved closer to the aperture	the picture size will [ ] increase [ ] decrease [ ] remain unchanged
e)	If the aperture is made larger	the picture will become: (mark all correct items) [ ] larger [ ] smaller [ ] fuzzier [ ] sharper [ ] brighter [ ] dimmer

Chapter 7

VARIATION OF LIGHT INTENSITY WITH ANGLE AND DISTANCE

With real world applications to temperature variations on Earth by location and time of year

This chapter first poses the question of whether light carries energy, and then tackles two further questions related to straight line light travel, namely: i. Does illumination intensity on a surface depend on the angle of illumination, and if so why? ii. Does the intensity of light depend on distance away from a point source, and if so why?

We explore both effects, and use our ray model of light behavior to explain them

These issues may seem at first to be rather formal or 'dry' aspects of light behavior, but it turns out that they have enormous real-life consequences! In fact dramatic consequences for all life on earth – or anywhere else in the universe. This is because the energy that the earth receives, as a planet, comes as radiation from the sun.

In this chapter we will first concentrate on the effect of illumination angle, then on the effect of distance, and finally look at the application of these ideas to explain planetary temperatures and the seasons.

SECTIONS

7–1   Does light carry energy?

7–2   Illumination at an angle

7–3   Variation of intensity with distance from a point source

7–4   Real world applications of the effects: implications for global temperatures and the seasons.

7-1 DOES LIGHT CARRY ENERGY?

We have studied several basic properties of light and how it travels. We now ask a further question: Does *light carry energy?*

Exploring

Can you think of any simple ways of testing whether light carries energy? Make some suggestions before reading on.

Some simple observations to try

- *Light from a light bulb.* Hold your hand near a light bulb – feel the effect and report.

- *Light from the sun:* Effect of sunlight on the ground or on your hand. To do this, you can compare ground in sunlight and in shade. Any difference in temperature?

- *Concentrating the light from the sun.* To make the energy effect clearer, we can concentrate the light from the sun by using a magnifying glass lens. Go outside, use the lens to the sun's rays down to a small spot, and see if you can make a burn mark on paper or wood. (Do you think one could start a fire this way?).

- *Light 'paddlewheel'.* The instructor may demonstrate.

- *Solar panels and solar cells.* These are used to convert light energy to heat or to electricity.

Conclusions

We have added another property to our knowledge of light characteristics!

Notes

Light travels vast distances, across empty space, e.g. from the sun to the earth, and carries energy with it. Somehow!

This fact has important implications. For example since light from the sun carries energy to earth, could this not be a major factor in earth's temperature and climate?

We now know that besides *visible* light from the sun there is also invisible radiation, also carrying energy.

7-2 ILLUMINATION AT AN ANGLE

A. Introduction and central question

Light may strike a surface perpendicularly, or at various angles. If the angle changes, what happens to the intensity of illumination on the surface? Does it change? How and why? Can our previous ray model of light handle this question?

Our central question for this section is thus:

How does angle affect intensity of illumination? And why?

We start by exploring and observing the effects of illumination angle, and then try to understand and explain them.

B. Exploring the effect of angle

We take a simple situation first (as always!). We will set up a beam of light, have it strike a surface, observe the illumination, then vary the angle of incidence and see what happens.

Setup

We need a light beam of limited extent. The beam could be from a flashlight, or even better, the square beam produces by an overhead projector when a card with a square hole cutout is placed on the projector surface. Set this up, and place a white board in the beam.

Pic of beam and surface

Observations and findings

First hold the board so the beam hits it 'head-on', and observe the illuminated region and its brightness. Then angle the board so the beam hits at a more glancing angle. In each case, focus on the following questions:

- How do the *areas* illuminated by the beam spot compare, for the head on and angled cases?

- How do the illumination *brightnesses* on the board compare, for the head on and angled cases?

What do you observe?

Wondering

We see that the brightness certainly changes as we change the angle. But why? Why should it be different for different angles? Can our existing knowledge of light help us to explain?

We can get a clue from the fact that brightness and illuminated area both change when the angle varies. The illumination gets dimmer, while the area illuminated gets bigger.

C. Analyzing using diagrams

To analyze this properly, we consider a parallel beam of light striking a surface and try to understand what's going on.

Draw a limited beam of light and let it strike a large surface head-on, making a bright illuminated area as shown in the figure. Then draw for a different angle, as on the right.

<u>Light striking head-on</u>   <u>Light striking angled</u>

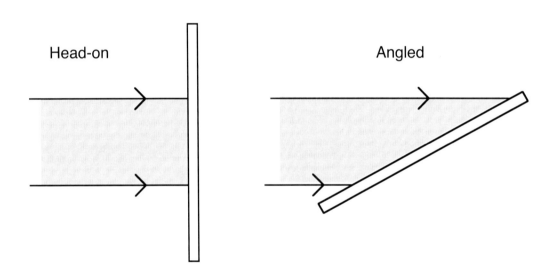

Head-on   Angled

In our hands-on exploring, we noticed that both the area illuminated and the brightness changed. Can we understand both of these effects, from the real setup and from the diagrams? We will start with a qualitative conceptual analysis, then make it quantitative.

Qualitative analysis

Comparing areas illuminated. Is it clear to you, both from real observation and from diagrams, that the illuminated area is greater in the angled case? The diagram helps explain why.

Comparing illumination intensities The same amount of light now has to illuminate a bigger area. From this can you develop a reasoned argument that the illumination brightness or intensity will be less? Discuss and formulate.

Conceptual understanding in terms of ray model. We can also use our ray model of light to see why the intensity should be less in the angled case. Redraw the above diagrams and then add a set of parallel rays within the light beam. Compare the spacing of the rays striking the

164

surface in the two cases. Does this help show why the intensity is different in the two cases? Discuss.

At this stage, we have hopefully reached a qualitative conceptual understanding of what is going on, and how to think about it.

Quantitative analysis

Next we want to be more quantitative. To do this we have to be able to compare areas numerically, and we will need a more scientific definition of 'intensity' rather than using the subjective term 'brightness'.

a) Area comparison

We need to make a quantitative numerical comparison of areas illuminated. First, does this with the real thing: make a square beam (using the cutout on the overhead projector) and measure the comparative areas illuminated for the two angles.

Result:

Then do the same thing using diagrams: think about what happens to the *length* of the spot and the *width* of the spot, and why. Which of these changes, which does not? Then work out the ratio of areas in the two cases.

Result:

b) Intensity comparison

Now we turn our attention to comparative intensities. Well, we have one immediate problem here – we haven't yet defined what we *mean* by intensity! Not scientifically anyway – have just talked of our subjective perception of 'illumination brightness" by eye.

A way to tackle this is to note that the light beam carries *energy*. We could thus usefully define illumination intensity as the energy falling on each unit of surface area every second.

Now let's develop a quantitative argument, for the particular case where angle is such that area is a factor of *two* greater. The limited light beam carries a certain total amount of energy each second. Now if the same total energy has to illuminate *twice* the area, then the energy falling on each unit of area each second will be a *half* of what it was before. Saying the same thing another way: if the total energy in the beam is spread out over *twice* the previous area, the illumination intensity on any unit of surface area will be *halved*.

We have illustrated with a particular case, but of course for other angles of illumination, the areas illuminated will be different, and so the intensities will likewise be different. The problems below give some practice in this.

D. EXAMPLE PROBLEMS – to consolidate understanding

These will be worked on and discussed in class.

1. Angled at 60° versus 90°

For light coming in at a moderate slanted angle of 60°, how will the surface intensity compare with head-on 90° illumination? Use scale construction and measurement to compare lengths, areas and hence intensities.

2. What angle will give two-thirds intensity?

At what angle will the surface illumination intensity be two-thirds that of the head-on case? Do by construction on a diagram.

3. Ground intensities compared for Miami and Kalamazoo

In Miami in January the sun reaches a maximum elevation of 51° at noon, while in Kalamazoo at the same time it only reaches 24° above the horizon. (Had you noticed that it never got very high?) Compare the intensities of sunlight on the ground at the two locations. Use ray diagrams to illustrate and compare the situations and to solve the problem.

Do you think this situation relates to the comparative temperatures in Miami and Kalamazoo in January?

7-3 VARIATION OF INTENSITY WITH DISTANCE FROM A POINT SOURCE

– Another consequence of straight-line light travel!

A. INTRODUCTION AND FOCUS QUESTION

We already know from an early chapter that light illumination diminishes further away from a point source. We now investigate this effect in more detail, first to understand it conceptually, and then to develop a quantitative law, stating exactly how intensity varies with distance.

The goals of this section are:

Firstly, to observe how light illumination changes with distance from a point source, propose competing theories about why this might be so, then test them out and select the best.

Secondly, to investigate the distance effect quantitatively, find how the area illuminated changes with distance, and propose a law for this.

B. OBSERVING THE EFFECT OF DISTANCE

Developing knowledge

Let's first observe the distance effect, then try to understand it.

Use a small birthday candle* flame as the 'point' source. (A piece of paper with a hole in it can serve as a 'stand'). In a darkened room, move a card nearer and further away from the candle, and note how the illumination intensity on it changes.

> *Note: we use a candle not a maglite bulb. A bulb gives undesirable streaky effects on the screen due to the glass of the bulb. And a candle if fun and can be used at home (with care).

Pic

Observations:

C. WONDERING WHY: SOME COMPETING IDEAS

As curious would-be-scientists, we naturally ask: *WHY*? Why does light seem to get dimmer further away? What's going on? We would like to *understand* this behavior, in terms of a physical reason for the illumination decreasing.

Let's start by gathering some competing ideas about this! What could be happening here? What alternative reasons can you think of? Brainstorm with others, even if an idea sounds a bit crazy, and make a list of possibilities.

Ideas   Source of idea?   Any evidence yet for idea?   Testable idea? How?

Here are some ideas, to add to your own: Does light maybe get absorbed by the air as it travels? Or does light gradually 'disappear' somehow? Or is it simply that the diverging light has spread out more? Maybe light gets 'tired' as it goes! (Or more scientifically, loses energy or uses up energy). Or what??

By now you will have gathered some tentative theories as *hypotheses.* We can even name them – the "Absorption" theory, the "Gradually Disappearing' theory, the "Spreading Out" theory, the "Energy Loss" theory, and any others that were proposed.

Our competing hypotheses – by name!

Now we can start exploring them further. Any idea how you could check out these competing ideas? That is, set up a fairly simple test to choose the most promising or rule out others?

CHECKING SOME OF THIS OUT - COMPETING THEORIES GO HEAD TO HEAD!

Being sensibly lazy, let's focus first on the idea that's the *easiest* to check out! :). That is consider whether the effect could be due to 'spreading out' of light. That is, is the effect due to the divergence, rather than any other reason?

How could we check this out? One way to compare hypotheses, i.e. to check out whether the dimming is due to spreading out, or another reason like being absorbed or disappearing, is to compare a spreading light beam with one that is *not* spreading, e.g. a laser beam.

Thus, to check out whether divergence is a reason, we need to see what happens for a beam which is *not* diverging! We can try a laser
168

beam, which hardly spreads out at all with distance. That's very different from the diverging light from a point source. Do you think that non-spreading laser light would also get weaker further away, or not? Doing this comparison would surely be a good way of testing which of our competing theories survives!

So, lets try it! Get a laser and a card. As you move the card further away from the laser, see if the intensity of illumination still varies as dramatically as for the point source. (Test over the same range of distances you used for the point source, for proper comparison). Note that this experiment also checks out the 'absorption' theory – if light is being absorbed by the air, the laser beam should also be absorbed, and get dimmer.

Test and result:

So, what can we now conclude? For a point source, what must be the reason that illumination intensity gets less when we go further away?

D. STUDYING A DIVERGING LIGHT BEAM CAREFULLY

Let's get professional now, and set up a clearly defined diverging beam of light to experiment with.

Setup: We can select a portion of the diverging light using a hole in a card. Cut a hole 1 inch square in a card (and keep the cutout piece, for use later). Position the hole about 8 in from a bare maglite bulb (point source), and place a screen beyond. The square hole lets a square diverging beam of light through, to experiment with. What do you see on the screen?

Pic

Qualitative exploring first

Start with the screen right up against the hole, and then slowly move it further and further away, noting what happens to both the **area** of the bright patch and its **brightness**.

(Dem).

What happens to area? What happens to brightness?

Does this make physical sense? I.e. that the brightness decreases as the distance increases? Think and discuss before reading further, and come up with your own explanation.

Is the following reasoning logical to you? The diverging beam carries a certain amount of light energy each second. If we go further away, this same amount of energy has to illuminate a greater area. So it is not surprising that the illumination becomes dimmer. To put it another way, there are more units of area to illuminate further away, so the amount of energy on each unit of area will be corresponding less.

Now for the quantitative

Thus far have explored qualitative behavior, of both area and brightness. But science is not satisfied with that, it wants something quantitative as well, i.e. numbers. That means a quantitative investigation. To make things simple, we can start by *doubling* the distance. We first look at the effect on area, then intensity.

Focus question: when we *double* the distance from the point source, what exactly will happen to the area illuminated?

To explore this, set up the point source and square hole, to get a square beam. You can set this up on a whiteboard. Put the screen near the hole, mark the outline of the square bright spot, then move the screen to twice the distance and mark the new bright area. Then compare the two areas. It is best to make a cutout of the first bright area, and overlay it on the second, in order to make a clear physical comparison. Can you think of a clever way to *compare* the areas, without actually measuring either one?

> You can move the screen alongside a meter stick on the whiteboard, for easy distance comparisons.

What do you find? Does this make sense?

Hence, what is the ratio of areas? . . . .

We have achieved part of our goal: we now know exactly what happens to the area illuminated by the beam when you double the distance!

Diagrammatic representation

We can show how this works on a diagram too. Draw a three-dimensional perspective diagram of the situation, showing suitable light rays and areas. Instructor will discuss.

Ray diagram (3-D perspective) showing light rays and area comparison. The instructor may also refer to books e.g. PSSC Physics.

On your perspective diagram, show how the *lengths*, *widths* and *areas* of the two squares compare.

Does the area comparison make sense in relation to the length and width comparisons? Explain.

Next: what if we *treble* the distance?

If we now put the screen at *three* times the original, what happens to area, and what happens to intensity?

Surely the same kind of thinking and argument will work for three times the distance? Go for it! Do this by hands-on experimenting, and by drawing the situation. Think about areas and energy etc. What happens to area? What happens to intensity?

Then of course we could do the same for <u>four</u> times the distance, and so on.

Hmm: Is there a pattern emerging?

Hence, propose general relationships i. between area covered and distance, and ii. between intensity and distance.

Radiant heat

We now know that light behaves like this, getting less intense with distance as it spreads out from a small source. So other things also behave this way, e.g. what about radiant heat from a small hot source? IS there also an intensity falloff with distance? Check it out as follows. A light bulb gives out both light and heat. You can feel the heat by holding your hand near to the bulb.

Sketch:

Then move your hand away a bit – what happens to your sensation of heat? Report and explain, in terms of energy.

Note for later: It turns out that light and 'radiant heat' are closely related – in fact they are different members of the same family! [The 'electromagnetic wave' family]. So it is not surprising that they should behave the same way in many respects.

E. REFLECTING ON KNOWLEDGE - CONTENT

· Note that the reduction in intensity is a direct consequence of the fact that the light is *diverging*, i.e. spreading out, from the point source. It's the divergence that leads to it having to cover a larger area at greater distance. Since the available energy has to serve a larger area, the energy on each single unit of area will be less. To phrase the idea differently, the energy gets more 'spread out' as the light diverges.

· An interesting realization is that our ray model for light behavior from a point source already contains within it the answer to the intensity-distance question! Note how this works if we simply show diverging light rays, in three dimensions, along with the idea of intensity as energy on a unit of area.

· Note that the 'falloff' in intensity in this manner will only be true for diverging rays. A laser beam, where the rays are parallel, is a different matter.

Approach - Contrast with an already-made treatment

Note that we have approached this section as science-in-the-making, posing questions, exploring, and building an explanation from evidence and what we already know about light. How might a conventional textbook present the concept of intensity and the relation between intensity and distance?

Either look one up, or imagine for yourself how a book might do it! Write down how these might be presented concisely and efficiently.

Found or created examples.

F. EXAMPLE PROBLEMS

Example problems

The instructor will provide example problems, formulated either in terms of ratios or values.

Homework Problem

How does the intensity of sunlight received by the planets Earth, Jupiter and Neptune compare? To do this, first find the distances of these planets from the Sun. (Either in a book or on the internet). Find the ratio of their distances, and round off to the nearest whole number. Hence make a rough numerical comparison of the intensity of sunlight at the three planets. Which planet would you expect to be coldest? Why? (Note: the planetary atmospheres will also affect the actual intensity received on the ground).

G. PROBING PEOPLES' EVERYDAY IDEAS ABOUT INTENSITY AND DISTANCE

Peoples' everyday ideas affect learning. So it will be of interest to investigate a few peoples; conceptions about why light and heat intensity fall off as you get further away. To do so, first demonstrate the effect to them, so they can experience it, and then ask for their ideas on why it is so, i.e. how it works. Do not 'lead' them or teach them, but just probe their ideas and thinking. Report back.

7-4 REAL WORLD APPLICATIONS OF THE EFFECTS

• Planetary temperatures   • Global temperatures   • The seasons.

We now know that light intensity varies with angle of illumination, and with distance from a point source. These effects have important real-life consequences, for our planetary system and for life on earth, as we will see. This is because radiation from the sun is the source of energy.

There are three different applications we will consider:

1. Comparing temperatures on various planets, e.g. Mercury, Venus, Earth, Mars, Saturn, etc.

2. Comparing temperatures at different places on the globe, e.g. Mexico and Canada.

3. Comparing temperatures in summer and winter, e.g. Minneapolis in July and January.

A. TEMPERATURES ON DIFFERENT PLANETS

Assuming that planets in the solar system most of their energy from the sun's radiation, how would you expect their temperatures to compare? Clearly, the variation of radiation intensity with distance can explain this.

We have already done one example problem on this (see earlier section).

For further problems, you can look up distances of other planets, and their temperatures, and see that the pattern makes sense. You can also calculate the ratio of energy received on these planets to that on earth.

B. GLOBAL TEMPERATURES AT VARIOUS LOCATIONS

We start with a real-life observation and question:

It is colder in Canada than in Mexico.   Why?

Let's gather possible ideas, i.e. brainstorm.

POSSIBLE IDEAS Yours and others	SOURCE OF IDEA	DISCUSS • True fact? • Valid reason? • Needs investigation?

(Note that respectable initial ideas don't have to be right, but they should be *testable* in principle, and based on evidence, else they are not really meaningful).

This is a particular example of a broader question, i.e. why do different locations on earth have different temperatures, even at the same time of year?

1. Viewpoint from Earth: observing the Sun

Before discussing the ideas further, let's look for *observations* we can make, which may help our thinking. We assume the earth is heated by radiation from the sun, and this affects temperature. So it makes sense to observe the sun, its location and daily behavior, at different global locations, and see if there is any difference.

We ask: what's different about the sun's location at noon, as observed from Canada and Mexico? And can this help explain the different temperatures? How? Discuss.

From your discussion, what might be the best explanation for the temperature difference at the two locations? Illustrate with sketches, showing the suns rays coming in to the ground.

2. Viewpoint from space: how sunlight strikes the globe

Another viewpoint we can take is the view from space, where we imagine looking at the earth and sun from the 'outside'. Let's see if this view can also help explain why one location on the globe is colder than another.

Look at an earth globe, and imagine rays arriving at it from a very distant sun. Compare the situations at the locations of Canada and Mexico. Anything here that can help account for the difference in temperatures?

3. Some questions about the situation

Is the light from the sun, arriving at the earth, pretty much the same whether it arrives at Canada or Mexico?

And if the rays from the sun are the same at both places, why should sunlight have less heating effect in one location than another?

176

4. A diagram showing sunlight striking two locations, from both earth and space viewpoints

We have considered this from both the earth (observational) viewpoint and a space viewpoint. It is interesting to illustrate both viewpoints together in one diagram, as shown below.

We will develop diagrams that compare the situations at Kalamazoo and Miami.

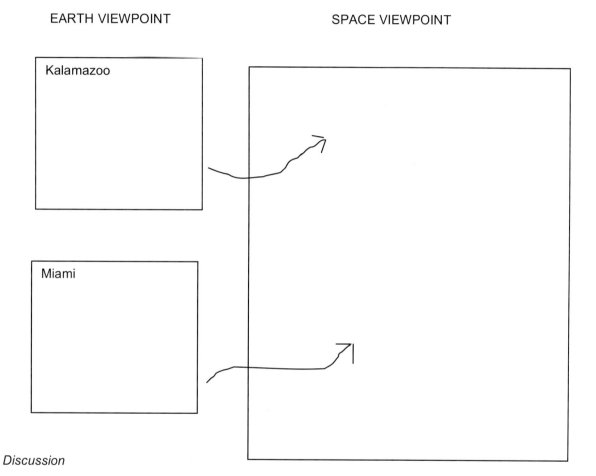

EARTH VIEWPOINT

SPACE VIEWPOINT

Kalamazoo

Miami

Discussion

From the diagrams, from both viewpoints, one of the reasons for different temperatures at Kalamazoo and Miami should be clear.

Remember however (from earlier discussion) that there is another factor operating to influence temperature; what is it?

C. "REASONS FOR THE SEASONS"

1. Question

Why is it colder in January than in July, in Kalamazoo?

Or the more general question:

Why is there winter and summer? What are the reasons for the seasons?

2. Some initial ideas – brainstorming

Possible ideas (yours and others)   Source of idea   Comment

Can our knowledge about angled illumination help explain seasonal climate variation? What other knowledge about the earth and sun do we need? Our assumption will be that the incoming radiation from the sun is what provides energy to heat the earth's surface.

This issue can be tackled from both the earth (observational) viewpoint and the space viewpoint.

3. Earth Viewpoint: Observational Evidence

We all see the path of the sun across the sky every day, at various times of the year. Have you paid much attention, observed the path and angle of the sun, and noted if it changes at all during the course of the year?

Sketch and describe the path of sun across the sky during the day, in winter and in summer, in Kalamazoo.

Sketches >

Might these actual observations be able to account for the fact that Kalamazoo is colder in January than July?

Note there are *two* aspects to consider, sun elevation and duration of day.

4. Space viewpoint: the sun illuminating the tilted orbiting earth

Demonstration of how the tilted glove orbits the sun.

Kalamazoo 'leans toward' the sun in July, and 'leans away' in January.

Imagine yourself located at Kalamazoo. Can you see on the globe how the angle of sunlight striking the ground will be different six months apart, when the earth is at opposite ends of its orbit round the sun?

The angle of incoming sunlight is different, hence the energy received on each unit of ground area.

See diagram on next page

5. How the seasons arise – a diagram showing both earth and space viewpoints

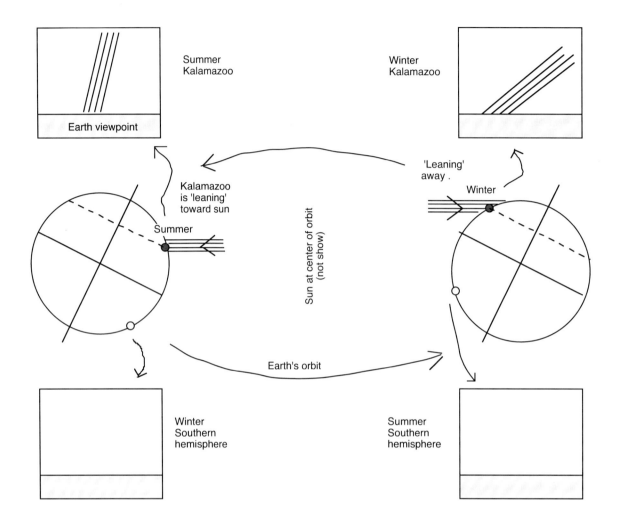

Summer
Kalamazoo

Earth viewpoint

Winter
Kalamazoo

Kalamazoo
is 'leaning'
toward sun

Summer

'Leaning'
away .

Winter

Sun at center of orbit
(not show)

Earth's orbit

Winter
Southern
hemisphere

Summer
Southern
hemisphere

6. Discussions

Instructor to lead.

Note that there are *two* reasons why it is colder in January than in July in Kalamazoo. In January the sun's rays come in at a more slanting angle to the ground, and also daylight is shorter. Both of these effects can be demonstrated in either the earth or space viewpoints.

If we look instead at a town in the *Southern* Hemisphere, the diagrams can equally well show the situation there. Note that it will be summer there when it is winter in Kalamazoo, and vice versa. Complete the diagrams above to show the southern hemisphere situation.

7. VIDEO "A PRIVATE UNIVERSE"

In the video *A Private Universe,* Harvard graduates are interviewed at commencement, and asked why the moon shows different phases, and why it is colder in winter than summer. The same questions are then asked of middle school students, both before and after instruction.

The list below suggests things to concentrate on, and questions to keep in mind while viewing the video.

A private universe: video viewing guide

In this 18-minute video people are asked about the seasons and the phases of the moon. The questions posed are:

    a) What causes the seasons?
    b) What causes the phases of the moon?

There are three sets of interviews on these topics: i. Harvard students on graduation, ii. 9th grade students before instruction, and iii. 9th grade again after instruction.

We will analyze what students say, what ideas they have, how they are thinking, where their ideas might come from, what kind of instruction they had, etc.

Some questions to keep in mind while viewing. Make notes as necessary.

- What are the common beliefs about the causes of the seasons?

- What reasons are commonly given for the phases of the moon?

- In responding, do people seem to be retrieving facts they recall from instruction? Or are they thinking it out? Or working from some principles?

- Do they state these their answer with assurance, whether right or wrong? Or are they tentative and thoughtful? Do any say they don't know? Do any of them seem to be just guessing?

- Lets' consider one 9th grade student, Heather, one of the best students. How does she answer? As if she knows what she's talking about? With some confidence or is she insecure? Does she say she isn't sure, or doesn't know? Does she seem to be working the answers out? From an understanding of the basics of how the system works?

- Where did that strange orbit shape come from that Heather drew?

- Why would a student bring in something she did not understand and where it was hard to see how it answered the question?   Does school educational dynamics lead to this?

- What do you think of the teacher, teaching design, quality and approach in this lesson?

- The teacher seems surprised at her students (wrong) responses and explanations. Yet she is in contact with these students in class every day. What does this mean about classroom instruction?

- The usual textbook 'perspective' drawing of the earth's orbit is deliberately done for a good intended purpose. Yet it seems to produce a misconception that the distance of the earth from the sun varies greatly. Is this an example of educational hazards? What to do? Discuss.

- The teacher uses three-dimensional models, with balls to represent sun, earth and moon. Is such a model demonstration a good way to teach about these phenomena? Or are suitable diagrams just as good, or even better in some cases?   What do we hope learners will learn to do? Discuss.

- Note that no *observational* aspects (of the skies, sun or moon) from the earth viewpoint are ever mentioned, by students, teacher or the video commentary. What do you think of this?

- The length of day never seems to be mentioned as a reason. What do you think of this?

- The responses show common misconceptions, which seem to occur no matter what level of instruction students have experienced. How can it be that 9th graders with no prior instruction in science, and graduates and faculty from Harvard, give much the same explanations for the seasons and phases of the moon?

- The prevalence of wrong ideas indicates that conventional instruction fails all too often. What is it about the teaching and learning of these topics that is not succeeding?

- Consider the way these topics are treated in our course. To what extent do you think the approach might help avoid these misconceptions about the moon and the seasons? Why?
- Etc. Add your own notes and questions.

Discussions

Reflections on knowledge

Note that from the earth (observational) viewpoint, the *same* two factors, acting together, give rise to temperature variation with location, and temperature variation with time of year. What are these two factors?

As to the daylight duration factor, variations arise because the axis about which the earth spins is tilted with respect to the sun's rays. You can most easily see this on a globe model of the earth as you slowly turn it.

As to the angle factor, this would work for variation of climate with location, whether or not the earth's axis was tilted.

Would there be seasons if there were no tilt? Explain.

Resource for discussion & ideas

Instructor may refer to Physlrnr archives of September 2003 on Private Universe and Seasons,.

Chapter 8

WHAT CAN HAPPEN WHEN LIGHT ENCOUNTERS MATTER?

In the course so far, we have explored the behavior of light, noted some of its properties, and devised a model. Thus we know something about light *emission and travel.* A natural next question to ask is – *what can happen when light encounters matter?*

This short chapter is a brief look at some of the most common things that happen when light interacts with matter. We explore this quickly here, discover the possibilities, and then investigate some of these more thoroughly in the chapters that follow.

This chapter is thus simply a 'prelude', setting the scene for the chapters to follow.

SECTIONS

8-1   Trying things out: light striking different types of matter

8-2   Observations: some different effects

8-3   Commentary

8-1 TRYING THINGS OUT: LIGHT STRIKING VARIOUS TYPES OF MATTER

We need to set up a light beam to encounter various types of matter, in various situations, and observe what happens. This can be a class demonstration, with group suggestions and predictions.

First suggest various cases to try. Also think of cases from everyday experience, where light and matter interact. The instructor will also add to the list.

Ideas to try:

Scientific inquiry frame of mind

In studying light and its behavior when it encounters matter, we take a scientific inquiry frame of mind. We explore what happens and ask ourselves – what does this tell us about light, and about matter, and how can we start to understand what is happening?

8-2 OBSERVATIONS: SOME DIFFERENT EFFECTS

For each of the cases below, note the sample of matter to be used, speculate what might happen, when the light beam encounters it. Then we try it, note the findings and illustrate with sketches. It will also be interesting to let the light beam strike at different angles.

A table for recording observations is provided below, but make your own sheets to get space for more detail.

Case 1: Black card or cloth	Case 2: White card
? - Ideas arising from this observation?	? -
Case 3 Colored card	Case 4 Mirror at angle
? -	? -
Case 5 Clear glass - perpendicular	Case 6 Clear glass block. At angle
? -	? -
Case 7. Colored glass or plastic	Case 8. Scattering. Dust or water spray.
? -	? -

8-3 COMMENTARY

Puzzlement and curiosity?

Most of the things that happen when light encounters matter are so familiar from everyday experience that we hardly notice them of given them much conscious thought. Thus we see light striking these kinds of surfaces every day. Yet when you start to think about each of them, is it not mystifying? How amazing that light be absorbed on one material, reflect from another, apparently change color on encountering another, go right through a hard solid substance, and be bent in doing so, etc. How can light be doing these things, and what's going on? Puzzlement and curiosity seem the right reactions!

How and why

We have started exploring HOW light behaves when it strikes matter, and found a variety of cases, which is interesting in itself. Note however that we cannot yet answer the deeper question of WHY it behaves in those ways. What happens to light clearly depends somehow on the nature of the material and of the surface.

Note that in order to understand WHY light interacts with matter in these various ways, we would have to know about the *internal makeup* of the matter, i.e. about the particular atoms and molecules that make it up, and understand their properties and interactions with light. This is a whole new endeavor on the microscopic' level, and is a challenge beyond the scope of this course. We will limit ourselves to investigating the 'macroscopic' behavior of bulk matter and light, and seeking concepts and laws to describe this behavior.

Next

We will investigate some of the above phenomena in detail in the units that follow. In particular, we will have units on reflection and refraction.

Notes

Chapter 9

REFLECTION, MIRRORS AND IMAGES

There are two rather different purposes to this chapter.

First we are going to investigate what happens to a light beam when it strikes a mirror surface. That is, we will investigate the physical phenomenon of reflection, and from our observations will formulate a law for reflection.

You might think that knowing about the phenomenon is enough, but it turns out that because of reflection we can use mirrors to *re-direct* sets of light rays. They will then look like they are coming from someplace else. This gives rise to all sorts of interesting and important effects! We will be inventing the concept of *image* to understand what is going on.

So there are really two main parts to this chapter:

  A. First the basic *physical* phenomenon of reflection,

  B. Then an essentially *geometrical* section on how reflection of ray-sets gives rise to images.

Redirecting of light by reflection turns out to have important practical consequences: it forms the basis of many optical instruments.

SECTIONS

PROBLEMS

9-1 PRELUDE:
DEMONSTRATIONS INVOLVING REFLECTION BY MIRRORS

The instructor may introduce some lead-in demonstrations and visuals involving reflection, as a prelude to our systematic investigations. Observing some actual phenomena first is to generate interest and curiously and to provide a shared experience of the phenomena we are about to study. We have for example a number of light sources, plane (flat) mirrors, a vehicle rear-view mirror, pairs of mirrors, angled mirrors, a corner reflector, a periscope, a kaleidoscope, and more. We will look at what happens to a beam of light striking a mirror, and how things appear to us when we 'see them' in a mirror.

We will not try to *explain* these demonstrations immediately because we first need to learn more about the physics underlying them.

So we will start by exploring the basic phenomenon of reflection, with the aim of finding a law. Then we will develop the concept of image. And finally we will come back to these demonstrations, armed with the knowledge needed to understand them.

9-2 THE PHENOMENON OF REFLECTION
– seeking a law

A.   THE PHYSICAL PHENOMENON

Focus questions:

What happens when a light beam encounters a mirror surface?

And can we find a law or principle to describe this?

Here is a diagram of the situation. A narrow beam of light strikes a mirror surface at some angle as shown. What happens? And what if we change the angle?

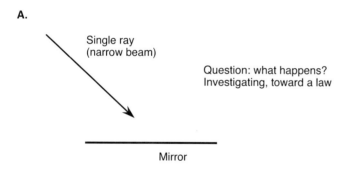

A.

Single ray
(narrow beam)

Question: what happens?
Investigating, toward a law

Mirror

We will investigate this and seek a law that describes it.

OUR APPROACH TO THE TOPIC

1. Developing knowledge. We will approach the phenomenon of reflection as science-in-the-making. That is, we will observe, pose questions, explore, and investigate experimentally, with the aim of finding a *law* for reflection

2. Testing and applying knowledge. We will then test our law and apply it to new situations and problems.

3. Reflecting on knowledge. We will reflect back on the topic and identify the essence and powerful ideas. We will consider how best to think and solve problems. We also consider how we learned these things and what might be good ways to teach

Procedure

We investigate this question experimentally in the following way: First we will observe and explore what happens, qualitatively. Then we will investigate, doing a quantitative experiment. Form our observations and data, we want to propose a law for reflection. We can then test the law further, and then use it to solve problems involving light reflection.

Note how we aim to get familiar with the phenomenon and its behavior qualitatively first, before taking quantitative measurements. This is good scientific practice. The explorations help us design a quantitative experiment.

1. Observing and exploring reflection behavior (qualitative)

This first stage is simply to observe what happens to a light beam when it encounters a mirror. Darken the room and use the ray maker and mirror. If this is done on a piece of paper or whiteboard you will be able to see the incident and reflected rays as the gaze the surface. Record what you observe.

Apparatus
Small plane strip mirror.
Ray-maker (e.g. maglite with slit-cap).
Protractor
Card
Screen

The next stage is to explore the *behavior of the phenomenon,* i.e. see what *happens* to the reflected ray as the angle of the incident ray is varied. Do this and note the behavior as the angle between beam and mirror changes, and illustrate with a rough sketch.

The instructor will follow up with a class demonstration and discussion of the effect and its behavior, probably by having students do it against the class whiteboard, again in a darkened room.

Discussion:

2. Designing an experiment

Design a quantitative experiment to find a law for reflection. This involves stating the aim, writing an initial plan of what you will do, how you will do it, and what data you will need to take. Be fairly specific, e.g. say what range of angles you intend to use, how many different readings, what you will measure, etc. Plan how you will record your data and draw up a table for your readings.

190

3. Experiment (quantitative)

Now carry out your experiment.

Take measurements and tabulate data.

Experimental record
Keep a full record of your experiment and results in your notebook.

Note on angle convention

In specifying the direction of a light ray, one could use either the angle between ray and surface, or between ray and the normal (perpendicular) to the surface. However scientific convention uses the latter, so we will too, but there is nothing fundamental about it.

Proposing a law

Analyze your results and propose a law for light reflection.

Proposed law of for light reflection:

Note. This law is yet another 'powerful idea' for light behavior, beyond the ones we developed in earlier chapters.

B. TESTING OUR REFLECTION LAW

Testing of our new knowledge – prediction challenges

We have investigated how light reflects, and proposed a law for reflection. But before we can feel confident about any law we should put it to the test. Can your law successfully predict the outcome of a new experiment?

So, let us test the proposed reflection law. At the same time, we can deepen our own understanding of how to apply the law.

The instructor will challenge you by setting up some experimental situations posed as 'prediction challenges'. For each situation, first predict the outcome using your reflection law. Then carry out the 'testing experiment' and see if nature gives the same outcome!

Testing experiment 1. – Predict the spot where a reflected light beam will strike a wall.

Lay your whiteboard on the bench. The instructor will draw a line along which a laser is to be pointed, and will place a mirror. Your task will be to place a 'target' on the wall where you predict the laser beam will hit after reflection. Show constructions on the whiteboard. (A paper plate marked with concentric circles makes a good target).

Once you have done this and placed the target, the instructor will set the laser along the original line, let it reflect off the mirror, and see if your prediction was reasonably correct. If the spot is close to the target this shows that your law of reflection works pretty well and you know how to apply it.

Testing experiment 2. – How to place a mirror so that the reflected beam hits a target spot?

Lay your whiteboard on the bench. The instructor will draw a line along which a laser is to be pointed, and will place a paper plate target somewhere on a wall of the room. Your task will be to position a mirror so that the laser beam, after reflection, will hit the target. Show constructions on the whiteboard.

Invent your own testing experiment

Try inventing your own testing experiment, to challenge other groups in the class.

C. SOME 'TOPIC KNOWLEDGE SUBASSEMBLIES' FOR REFLECTION

So far we have studied only the basic phenomenon, so a phenomenon diagram and a behavior (variation) diagram should be sufficient knowledge subassemblies.

Phenomenon diagram

Devise a diagram to represent the essence of the basic phenomenon, i.e. the reflection of a light ray.

Diagram:

[Check your diagram against this note. Both the phenomenon and the law are rather simple, so the diagram will be simple too; it can be a sketch containing the following aspects: reflecting surface, incident and reflected rays, with equal angles marked].

Behavior (variation) diagram

We can add another diagram to show the *behavior* of the reflected ray as the incident angle is varied. This would include say two or three different angles on the same diagram.

Sketch such a 'behavior' or variation diagram.

Diagram:

Note

Strictly speaking this second diagram is not essential, since it is *implied* in the first, but useful knowledge subassemblies should *explicitly* represent useful knowledge aspects, for quick mental access to those features. Similarly, the second diagram incorporates the first, but its complexity hides the simple phenomenon. For these reasons it is useful to have *two* diagrams, each with its own purpose.

Mental images

You can form mental images of these two diagrams, to access whenever you need to think about the *basic* reflection phenomenon and its behavior.

D. APPLICATION – *applying* your reflection law to problems

Now that we have developed a basic law for reflection, we can start working with it. It is a 'powerful idea' which we can *use* to understand new situations, solve problems and see it's practical uses.

Problem: Reflection by two mirrors at right angles

A room has two mirrors on adjacent walls as shown (top view of room). A laser pointer shines a narrow pencil of light toward one mirror as shown.

ROOM

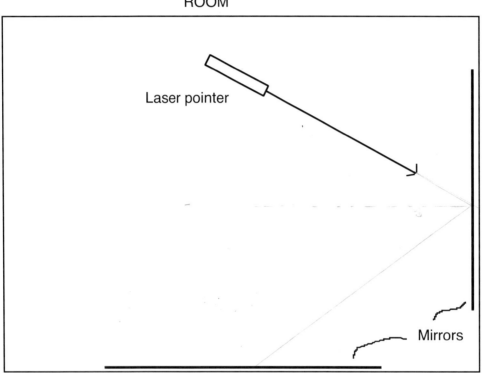

a) After the beam has reflected twice, what wall will it hit and where? Trace the path of the ray accurately through two reflections, and mark the final spot on the wall.

b) After the double reflection, how does the *direction* of travel of the light beam compare with its original direction? By inspection, does it seem to be parallel? Or not? Same direction? Opposite direction? Is this coincidence? To check, try a ray coming in at a somewhat different angle and see what happens.

c) *Prove* that after two reflections by mirrors at right angles, the emerging light ray will be parallel to the original incoming ray. This will need some geometry. [Optional question].

Note: Corner reflectors

This arrangement of two mirrors at right angles is called a 'corner reflector'. It has the interesting property of exactly reversing the direction of an incoming light beam – reflecting it back in the opposite direction – no matter what direction it comes from.

Three mirrors forming the corner of a cube will do the same thing in three dimensions, and is the basis for the 'cat's eye' reflectors used as lane markers along roadways. They reflect car headlight beams right back to the car, so the driver sees them as 'cat's eyes' shining brightly by reflected light.

194

Visualizing the basic physics in operation:
Ride a ray and obey the law!

Remember that the basic physics is the important thing, i.e. how a ray travels and reflects. For teaching young students there is a useful way to visualize this that 'puts them in the picture'. You can think of yourself as "riding a light ray" in a straight line toward the mirror, and then "obeying the law" at the mirror, and then continuing riding straight in the reflected direction. Make a cartoon sketch of a stick figure riding a light ray.

Riding a ray:

Further problems

These are at the end of the chapter. They will tell you when you really understand.

Next

At this stage we know exactly what happens to a narrow beam of light when it strikes a mirror, and have a law for it,

But in optics the topic of reflection involves more then just how an individual ray of light reflects. In real life we know from experience that we can see whole realistic scenes "in a mirror", while realizing that they are not really there, but elsewhere. How does this work? That's the task of the next section!

9-3 MIRRORS, RAY-SETS AND THE CONCEPT OF *IMAGE*

A. INTRODUCTION AND CONTEXT

From the last section, we found out how a narrow beam of light reflects, and formulated a law for that basic phenomenon. Well, it is one thing to know exactly what happens to a ray of light when it strikes a mirror, but in real life we have the complex situation of extended objects (like chairs and people) giving off light in all directions from every point of their surfaces; if all these various light rays strike a mirror surface and are reflected, then what will be the overall effect? That is, what will we see if we look toward the mirror and receive all this light? This is the situation we want to explore next. To do this we need to look at (so to speak) what happens to spreading bundles of rays from sources, when they encounter mirrors. This will lead to interesting 'mirror' effects – which we are so familiar with from our everyday experience that we may not give it much thought. We simply 'know' we can somehow 'see' things in a mirror. But how does this work? We want to explain and predict these mirror effects – with explanations based on our fundamental law of reflection of course. We won't need any more fundamental physics than this, but we surely will be using lots of geometrical diagrams as we trace what happens to sets of light rays! Doing this will lead us on to invent the concept of "image".

But before we introduce mirrors and images into the picture, lets first look just at the source situation: rays diverging from points on a source or object, at our eyes receiving those rays, and what we see. Once we understand this well, we can add the mirror and go from there.

B. HOW WE 'LOCATE' OBJECTS VISUALLY:

Before we go on to deal with 'seeing things in a mirror', let's first think hard about how we see things anyway, and how we locate where a light source is, from the light rays it omits.

You already know from an earlier chapter how "seeing" works generally and the role of light and the eye. Now it time to extend this a little – how do we know exactly where a source of light is located, from the light which reaches our eyes? That is, when you look at an object, seeing it by the light it emits, *how do you judge where it is?* I.e. in what direction and how far away?

Perceiving just the *direction* of an object is no problem – you see it as being in the direction from which rays are coming to your eye. But that doesn't tell you how far away it is – rays come from that direction no matter whether the object is near or far. So how then do we judge *distance*?

Judging distance:

i. Try judging distance with one eye

How do you judge how far away an object is when you look at it? It turns out that two eyes are much better than one for this! Here is an exercise: Have a friend hold out a finger horizontally. Close one eye, move your hand in sideways and try to touch their fingertip with your own. What happens? Do you hit or miss?

Now do the same using *two* eyes. Any more success?

This exercise gives us a clue about how we judge distance visually. Ideas?

Sketch of situation

ii. Locating objects visually with two eyes

If you think about that is happening, the way we see and locate an object is by viewing a *set* of diverging light rays coming from it and entering the eyes.   Three rays coming from a point source are shown.

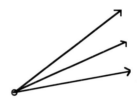

• Why is the *divergence* aspect important for locating the source?

• Why are at least *two* eyes (ray receivers) important for locating?

Is it clear how the origin of the diverging rays gives us the source location?   Note that a *minimum of two* diverging rays is needed, though we show three above.   Note also that it doesn't necessarily have to be your own eyes that receive the diverging rays.   It could be any two receivers, for example two people at a suitable distance apart.

Diverging Ray-Sets – "DivRaySets" for short

The idea of a **set of diverging rays** will be important in optics, since it represents how light is emitted from each point of an object, and shows where the light is coming from.

Understanding how we locate light sources visually, by receiving rays, helps to appreciate why we are giving special consideration to "diverging ray-sets'". Their role will become even more important when we introduce the idea of 'image' and need to decide where images *appear* to be located. We invent our own name for a diverging ray set – we call it "DivRaySet" for short.

Often we will work with just two rays, and imagine the rest.

The minimum diverging ray set - two rays

External material on how we locate objects visually

The textbook 'PSSC Physics' has a nice treatment of this: the instructor can provide it separately.

C. "SEEING THINGS IN A MIRROR" - a practical look first

We now know that when we 'see' an object as being at a particular location, it is because rays of light are diverging from that location and entering our eyes.   That seems pretty straightforward. However, there is a more challenging situation: from your experience with *mirrors*, you will know that you can "see things in a mirror".   That is, if you look toward a mirror, you can see objects "in the mirror", when you know the objects are not really there at all, but somewhere else.

Try it: look toward a mirror on the wall: you will 'see' things that are actually elsewhere in the room.   (Including yourself!)

And this mirror effect is pretty realistic! We are used to mirrors from lots of experience with them, but young animals are not; so when a kitten sees "itself" in a mirror for the first time, what is sees is realistic enough that it believes there is another kitten there! It hisses and arches its back, and even goes around the mirror to look for the intruder! [URL to video]. So, what we 'see in a mirror' is certainly realistic, almost as if a real object were there.

So what is going on? We know how we see objects in the normal way (from diverging rays), but now, using a mirror, we see objects as being in locations where they are not! Is the eye being fooled? How does it happen that rays of light coming from the mirror are just as if they were coming from a real source? Sure, there is no real object actually there, but clearly the light rays coming to your eyes are the same as if there really were something there. And exactly where does it seem they are coming from? At the mirror? Behind the mirror? In front of the mirror?

Clearly what we are seeing must have to do with the reflecting of light from the objects off the mirror. And since we know the law of reflection we should be able to use our knowledge of reflection to explain the effect theoretically.

But before we do this, let's use actual mirrors to explore the effect practically, to sharpen up what we recall from experience.

a) Stick-Pointing Location Activity

Set up a bare light bulb as a source, and mount a mirror at about the same height on a wall.   Two students have long sticks with arrows on one en, to represent light rays.

1. First, to show how this works for locating the source itself, the two students look at the bulb from somewhat different places, then hold their stick along the line of sight, arrow pointing toward their eye.   It should be clear to everyone that where the lines of the sticks intersect is the source location.

Activity and notes:

2. Now the activity is repeated using the mirror. The two students look into the mirror from somewhat different positions so that each can see the bulb "in the mirror". Again, they then hold their sticks along the line of sight, with arrow end pointing to the eye.   Looking back along these 'rays', and seeing where they intersect, it should

be evident t to everyone in the room where the light rays 'appear' to be coming from after reflection. This large-scale demonstration is a convincing indication of where the "image" of the bulb t is located.

Where does the bulb image appear to be? On the mirror? Behind the mirror? In front of the mirror?

b) Locating by comparing with a real object behind the mirror

We can do another simple experiment to find the location of what you see in the mirror.

Lay a whiteboard on your bench. Place an object directly in front of an upright card. Mark the object and card on the whiteboard. Predict where an image will be seen when the card is replaced by a mirror, and place a 'marker' object at the predicted spot

Now replace the card by a mirror, to see if the image is indeed seen in the location predicted.

For this test it is most effective to use a "Miro' mirror (half-silvered semi-transparent mirror), or else a short mirror which you can see over the top of. In either case you should be able to see the image and the 'marker' object at the same time.

Is the image seen at the predicted place? (I.e. at the same place as the marker object). Is the result a surprise?

We have already learned enough physics about light and reflection to work out what is going on, so in the next sections we approach the situation systematically.

D. REFLECTION OF RAY-SETS
AND INVENTING THE CONCEPT OF IMAGE

Physics and geometry working hand in hand

We know that rays diverge from in all directions from a point source: here is a picture of a bundle of such rays. In an earlier section we noted that if one observes the rays, one can work back to where they originate, and thus one knows the source location.

Fine, we can tell from a diverging bundle where the light originates; but what if here is a *mirror in the way?* The rays directions are changed, and if we observe the reflected set of rays, they will appear to be diverging from somewhere else! It will appear to us just as if there were a light source at that new location.

How can we explain that theoretically using the law of reflection?

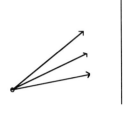

The situation of interest

Focus questions

- What happens when a diverging **set** of rays from a point source is reflected from a mirror?

- What optical effects arise from this 'manipulation' of a ray-set by reflection?

Start simple, work to more complex

As usual, we will start with the simplest situation, i.e. a plane (flat) mirror and a singe point source, to develop the basic procedures and concepts.

Then we will move on to *extended* sources of light. This is relevant in the real world because ordinary illuminated objects around us act as extended sources. The question will be: how is an image produced for an extended source, and what will it be like?

Then in the following chapter we will go on to curved mirrors.

E. TRACING DIVERGING RAY-SETS UPON REFLECTION

– where do the reflected rays they appear to be diverging from?

Introduction

To work out what happens when we have an object in front of a mirror, we simply have to consider what happens to all the rays of light coming from the object and encountering the mirror. To do so, we first consider what happens to rays from a single point source, since the object can be thought of as made up of lots of point sources.

Consider a set of many rays diverging from a point source – three of them are shown in the figure. From the divergence of the rays we can locate where the source is, simply by tracing the rays back to where they are diverging from.

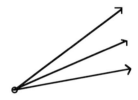

A diverging ray-set or 'DivRaySet' from a point source,

Fine, but our more interesting aim now is to find out what happens to this diverging ray-set (or 'DivRaySet' for short) upon reflection by a plane mirror. What will the set of rays look like now?

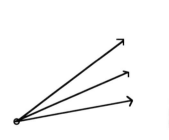

The DivRaySet going toward a mirror surface. What will happen to it ?

What will happen to the DivRaySet now?

Of course we already have the required tool to find this out – the law of reflection! We know how each ray will reflect, so we just apply the law to the rays one at a time, and see what *new* set of reflected rays results.

Example case

Let's try, on a larger diagram. The diagram below shows three (of many) rays diverging from a point light source and striking a mirror. Reflect the rays accurately, obtain the resulting ray-set and note where the reflected rays **seem** to be coming from. To do this you can extend each reflected ray backward using dashed lines.

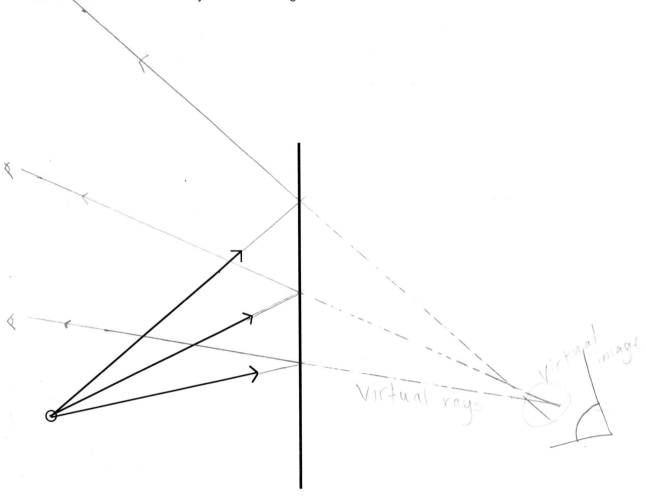

Note

We suggest drawing just two diverging rays first, rather than three. It should work either way, but its' hard to draw and reflect three rays accurately enough that all three reflected rays have a single intersection point, and we don't want side issues muddying the simple principle of how images originate. Later, add the third ray and see if it give a pretty consistent result. Of course with real light there would be innumerable rays doing this.

F. TRANSFORMED RAY-SETS, THE 'AS IF' PRINCIPLE, AND THE 'IMAGE" CONCEPT

The transformed ray-set

Look at your resulting ray-set after reflection – it is another diverging ray-set! It is a transformed set, produced by the action of the mirror. Note that this set appears to be diverging from a different point, not from the original source point. Mark the new point where the reflected rays *seem* to be coming from. Note that there is no real light source at that point! But rays seem to be coming from that location.

The "as if" principle

We see that the reflected ray-set looks "as if" it were diverging from a particular point behind the mirror, even there is no real source there, and no real rays exist behind the mirror. However to the eye receiving this diverging ray-set, it appears exactly the same "as if" there was a source of light at this location! In fact if the observer were unaware of the mirror arrangement, and just aware of the transformed rays entering the eye, she would think that she was looking directly at a real source. (Note: when children and animals first encounter a mirror, they tend to think they are seeing real objects, until they learn from experience. Have you seen a kitten hissing when seeing itself in a mirror?)

The concept of image

To deal with this visual effect, we introduce the concept of "image". We say that the mirror produces an optical 'image' at this location.

It is an image of the actual source, and arises from the reflecting action of the mirror, which changes the directions of the rays.

A point source gives rise to a point image, at a different location.

> *Questions*
>
> What is the minimum number of rays required in a diverging ray-set in order to locate an image?
>
> What if you tried to use just *one* ray – how far could you get in finding image location?

Demonstration of the imaging effect

We will demonstrate the 'imaging effect' using a flashlight and mirror.

i. First, set up a flashlight pointing toward a wall and produce a bright spot on the wall. Then hold a mirror in the beam, and produce a spot on a different wall. What is going on? If you were aware only of the second spot, you might think that a flashlight in another location must be pointing at it. We know that is not the case, yet the effect is the same "as if" there were such a flashlight pointing at it.

ii. Someone will turn on a bare light bulb at the front of the room. Note where you see this light source. Now look toward a spot at one side of the room – do you agree there is no light bulb at this location? Now someone at that spot holds up a large mirror and start adjusting its angle – as you watch, is there a situation where you seem to see the light bulb there, i.e. in that direction? If so, where does it seem to be located exactly? Is there really a light bulb there? Does is look "as if" there is? What are you seeing exactly? What role is the mirror playing?

Does seeing this demonstration and thinking about it help you to grasp the image concept?

The nature of the image

Looking back at the image formation diagram you drew, note that there are no actual light rays behind the mirror, so there are none at the location of the 'image'. All real light rays are in front of the mirror. The dotted lines that we drew to intersect at a point behind the mirror are simply 'extensions' of the diverging reflected rays. We draw the dotted lines to show where the reflected rays "appear" to originate from – an 'image' point. This kind of image, where there are no real light rays at that point, is called a "virtual" image. It would not be possible to put a screen at the virtual image location and 'capture' the image on screen, since there is no actual light there to illuminate the screen. To the eye however, what you see in the mirror looks like the real thing.

When we come to curved mirrors, we will see that another kind of image is also possible.

G. FORMULATING AN IMAGE-LOCATION RULE FOR PLANE MIRRORS

Focus question

By both experiment and ray construction, we found that the image in a plane mirror was located behind the mirror.

Our question now is: *exactly how far behind the mirror?*

Comparing distances by measurement

Let's check: How does the distance of the image from the mirror compare with the distance of the source from the mirror? Compare by actual measurement on your diagram.

Proposing a 'rule' for image location in a plane mirror

In summary, what do you conclude about image location for a plane mirror, and the distances of object and image from the mirror? Write this as a rule or powerful idea for plane mirrors.

Proving the image location 'equal distance' rule using geometry

Can you prove this 'equal distance' rule *geometrically*, using the triangles in your diagram?

Applicability of the equal distance rule

Note that we reached this conclusion about image distance for the special case of a plane mirror. We cannot automatically assume that the rule will generalize to curved mirrors; and will have to investigate this separately.

Notes and discussion: primary and secondary powerful ideas

We saw that plane mirrors give rise to 'images' when they reflect light rays, basically as a consequence of the law of reflection. The reflection law is a *fundamental* or *primary* powerful idea, which applies generally to any reflection situation. For the particular case of a plane mirror acting on diverging rays, we *applied* the primary powerful idea and *derived* another powerful idea, for the *location* of the image. We called this the plane-mirror image-location rule, or 'equal distance' rule. Note the status of this rule: it is a *secondary* powerful idea, because it can be derived from the primary idea plus geometry.

Thus we can say that not all powerful ideas are created equal! Some are fundamental and broad in scope; some are secondary and more limited in scope.

The image location rule, although it is secondary and case-specific, is pretty powerful nevertheless! It enables us to find images in plane mirrors quickly and efficiently. Much faster than if we had to go back to the fundamental law every time, trace at least two reflected rays and find where they appeared to diverge from.

H. TESTING OUR IMAGE-LOCATION RULE
– Experimental tests of the plane-mirror rule

Before using our image-location rule in practice, we should put it to experimental test in a few cases, to confirm that it works in reality! Some 'testing experiments' follow; one for use in class, one for formative assessment.

Image location test 1.

Object in front of mirror – predict the image location

(Note: this is the same experiment we used earlier as a prelude).

Lay a whiteboard on your bench. Place an object directly in front of an upright card. Mark the object and card on the whiteboard. Predict where an image will be seen when the card is replaced by a mirror, and place a 'marker' object at the predicted spot.

Now replace the card by a mirror, to see if the image is indeed seen in the location predicted.

For this test it is most effective to use a "Miro' mirror (half-silvered semi-transparent), or else a short mirror which you can see over the top of. In either case you should be able to see the image and the 'marker' object at the same time.

Is the image seen at the predicted place? (I.e. at the same place as the marker object).

If so, you can have confidence in the 'secondary' powerful idea about image location.

Image location test 2.

This will be provided by the instructor as formative assessment for learning.

It will also be used to discuss experiential intuitive reasoning and p-prims in optics.

Include the discussion in your notes

I. VARIATION BEHAVIOR:
WHAT IS THE EFFECT OF *VARYING* THE SOURCE DISTANCE? HOW DOES THE IMAGE BEHAVE?

So far we have done an image construction for just one particular source distance. What if we *vary* the source distance? That is, what will happen to the image if we move the source closer to the mirror, or further from the mirror?

We could of course draw new constructions for these cases, but maybe we can see what the behavior will be by looking at the ray diagram and *imagining how it would change* if the source distance changes.

Try this! You can do it in your imagination, or sketch on paper.

What do you conclude?

Note: there is more than one way to do it

Note that there are at least two ways to do the above. Either keep the ray set the same (i.e. keep the same divergence angle), or keep the place on the mirror the same where the rays strike (which means choosing a differently diverging ray set). The instructor can discuss. Which method did you use? If you like, try the other method also, and see how it gives the same result (as it must).

Note also that you could say how image distance behaves by simply using the equal-distance rule – as the source distance changes, so likewise does the image distance. However we are interested in seeing why the image distance behaves as it does, and for this a 'rule' is no use, we need to visualize the underlying principles and processes operating.

Idea power!

The nice thing about thinking about things in this conceptual way is that afterwards you will understand **why** the image behaves as it does as you vary the source distance. That is, you will have a conceptual grasp of what's going on, will be able to predict the effects, and will be able to explain them to others. And you should be able to do this anytime in the future, long after you have forgotten what actually happens.

Check your theoretical predictions against reality

You have predicted what will happen to the image by using the ray model; that is, this is a theoretical prediction. Of course you should also try the real thing. That is, set up a source and mirror, locate the image, then change the distance of source from mirror and see what happens. What happens to the image location – does it move closer, stay put, or move further back? Is this in accordance with what you worked out from theory?

Example problem

Instructor

Notes

COGNITION IN THIS AREA:
Topic essence subassemblies for images due to reflection

Here we will consider *cognition* in the area. That is, how best to think in understanding this topic, and how to think in tackling problems in the area. We will first re-introduce some of ideas already described in the first chapter, just for completeness.

Introduction

An understanding of any topic involves several related types of knowledge e.g. the basic phenomenon, its behavior, the principles, the procedures, the result features, and various cases. For thinking about a domain, it can be useful to represent some of this topic knowledge in a semi-compiled form.

In Chapter 1 we introduced useful kinds of 'topic essence subassemblies' that could represent the essential features of a topic and serve as tools for both knowledge retrieval and thinking. Various possibilities were phenomenon diagrams, behavior diagrams, principles & procedures diagrams, feature diagrams and case diagrams.

Now that we have studied *image formation due to refraction,* we will look back on the essence of the topic, and devise a number of useful subassemblies to encapsulate the knowledge we developed. (Previously, we had diagrams only for the basic refection phenomenon and its behavior).

Principles and procedures subassembly for image formation

The instructor will generate and discuss such a 'subassembly' with you, for the case of image production in a plane mirror. We try to include as many useful aspects as possible, e.g., the basic law, the procedures, the geometry, and the features arising.

Generate a 'knowledge subassembly' for the case of plane mirror imaging

Note that this subassembly incorporates a lot! Our basic ray model, the law of reflection, use of diverging ray sets, the concept of image, and the equal distance feature.

Ideally a topic knowledge subassembly will contain the essence of our understanding, i.e. the situation, the principles, the procedures, and the characteristic features. It will represent all this in a fairly **abstracted** way, so that it will be relevant to this **type** of situation whatever the surface details of particular cases.

Features subassembly

We will give an extended development of this, to model the idea.

Identifying 'characteristic features' for the plane-mirror image case

Although the image due to a plane mirror can always be located using *first principles*, i.e. by tracing reflected diverging rays (and you should always be able to do that), it is immensely useful to identify the "characteristic features" of plane mirror image situations.

The main features of the plane mirror case are that:

- The image is **behind** the mirror,
- It is at the **same distance** from the mirror as the source, and
- It is **virtual**.

These *features* follow from applying the law of reflection to a plane mirror geometry, so in that sense the features are secondary and limited to this geometry. But of course the features are interesting and useful in their own right, and it is of interest to picture them.

Features of cases can often be shown and visualized *pictorially*. A diagram can often represent features at a glance. Let us devise one for the plane-mirror image case. The instructor will develop it with you and discuss.

We can add text to feature diagrams to make them more complete and explanatory.

Features diagram for plane mirror image situation – showing image *features*.

Note that a features diagram is often tied to a particular case, but is still abstract and general enough that it could apply to a range of situations within this type.

The "why" behind features

Of course you should also understand *why* these features arise, in terms of the basics. Knowing the features and understanding the reasons behind them makes for deep understanding.

In other cases take care to check which features may or may not still apply

Of course since we identified these features from just one or two examples, we would be wise to check that the features do indeed generalize, by looking at some different examples also. Do this for yourself.

Thinking in terms of case features and in terms of principles

Thinking in terms of characteristic features is an efficient way of operating. It is a form of 'case-based reasoning', i.e. thinking based on cases you have already seen and understood. However do not just 'memorize' features. That is not adequate scientific thinking, and can lead to errors in new situations unless you also think in terms of basic principles, i.e. use 'principle-based reasoning' as well.

Behavior (variation) diagram

Case comparison diagram

Mental images

You can form mental images of these diagrams, to access whenever you need to think about images due to refraction and their properties and behavior.

Thinking back on knowledge so far

Identification of powerful ideas – primary and secondary principles

For reflection, mirrors and images there are only two *primary* powerful ideas or principles, namely:

1. Light reflects at the same angle it came in.

2. A reflected ray-set diverges "as if" it came from a particular location – the 'image' location.

There is also a *secondary* powerful idea for plane mirrors, namely:

3. The image produced by a plane mirror is the same distance behind the mirror as the object is in front.

Plane mirror image problems

Note that we can always do plane mirror image problems in two ways, either:

i. by going back to *basics* (the law of reflection and ray construction), or

ii. by using the *derived* properties of images, i.e. the image location rule.

You should aim to become proficient in both methods, i.e. be able to use both primary and secondary powerful ideas.

Notes

9-4 IMAGES OF *E X T E N D E D* SOURCES

Extended sources of light – will reflection produce an extended image?

Now that we understand imaging for point sources, we can turn to *e x t e n d e d* sources, An 'extended' source of light is one which has 'extension'; i.e. some size and shape. The source could be an active light emitter like a long fluorescent tube perhaps, or it could be an illuminated object. Very importantly in the real world, it includes ordinary objects like chairs or flowers or people etc., because these objects, if they are illuminated, act as secondary sources of light.

Our focus questions will now be: if light from an extended source is reflected by a mirror, will an image be produced, how is it produced, what will it be like, and where will it be located?

We might suspect that we can already work this out in principle, if we keep in mind that an extended source of light is just a collection of many point sources. After all, we know how to find the image of each of these point sources. Would not the image of the extended source just be the collection of these point images? Maybe we can use this 'insight' to proceed.

Reflecting back: why emphasize "DivRaySets" for imaging?

You may wonder: why have we been concentrating so much on *'diverging ray-sets'* from a point source? The answer is that **every** point of an **extended object** gives off a *ray-set of diverging light*. Thus if we want to know what a mirror does to light from an *extended* object, we can treat the object as a bunch of point sources each giving off its own DivRaySet.

The figure shows how various points of an illuminated object (potted plant) emit diverging light. A ray-set from each point, once reflected, gives rise to a corresponding image point somewhere else. So we can apply this to locate the images of *all* points of the object – and hence the image of the whole object.

Images of extended sources

Most objects in real life are extended. To understand what the corresponding image will be like, it makes sense to start by working it out for a very *simple* extended source.

Note: we will treat this as a *problem* – an extension-of-knowledge problem. Many conventional textbooks simply provide the method and conclusion, to be understood and learned, i.e. they do the thinking for the learner and present the result. Instead, we take it as a challenge to work it out for ourselves, since we have the underlying powerful ideas at hand. It will not be hard!

Let us start with the simplest kind of extended object. We choose an arrow – just a line with two distinct ends. (Why do we choose such a shape?) If light coming from this arrow encounters a mirror, what will be the effect? What image will we get, what will it be like, and where will it be?

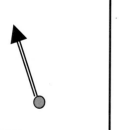

Note that from your work on point sources and images, you have *two* ways to tackle this – from primary or secondary principles. Use both. That is, first trace light rays using the primary law of reflection, then use the secondary image-location rule.

Work out a procedure with your group, and find the image of the arrow. Use **both** methods. Use a whiteboard or a large copy of the diagram above.

Then discuss:

Finally check the procedures you used against the following.

i. By first principles (ray tracing).

Trace two (or more) diverging rays from one point of the object (tip of arrow), reflect them from the mirror, see where the reflected rays appear to diverge from, and hence locate the image point. This gives the image of the arrow tip. Then we repeat the procedure for a source point on the base of the arrow, to get an image of the base. The in-between points will be imaged in-between (you can check this), and in this way we obtain the whole extended image of the entire arrow.

Note: the method depends on our earlier powerful idea about light sources, viz.: An extended source acts as an array of point sources, each of which emits light in all directions.

ii. By secondary principles (image-distance rule for plane mirrors).

Place the image point of the arrowhead behind the mirror, at the same distance as the source point is in front. Use the same procedure for the base of the arrow, and any in-between points, and hence get the complete extended image.

Did these two methods give the same result?

Note: Note how the source and image arrows are oriented – closer points imaged closer, farther points farther.

Note

There is more we could look into about images of extended sources, but we will not go into this here. Some of it we will leave for the problems. But just to mention one interesting point: you might have noticed something about things you see in a mirror –they look much the same as the real thing, but not quite; the mirror changes the way things seem to be facing, so that left and right seem reversed.

Can our simple law of reflection and the geometry of ray reflections really explain all this? Yes!

9-5 REFLECTING BACK ON THINGS (and a brief look ahead)

Light reflection – the physics and the geometrical consequences

We said at the beginning that there would be two different but inter-linked aspects to our study of light reflection.

The first was the *physics* of the phenomenon itself; here we investigated reflection and formulated a law.

The second was the *geometry* involved when light rays reflect from surfaces of various shapes; in particular we looked at the geometry of diverging sets of rays reflecting from a plane mirror. This led us to the concept of *image*.

We recap these two aspects of the topic pictorially below.

I. The physical phenomenon of reflection

Our focus question was: what happens to a ray of light encountering a mirror surface?

We investigated this experimentally by studying the behavior of a narrow beam of light striking a mirror surface, and found a *law* for reflection, namely:

. . . .

. . . .

. . . .

. . . .

A.

Single ray
(narrow beam)

Question: what happens?
Investigating, toward a law

Mirror

II. Reflection of a diverging ray-set, and the formation of images

A ray-set diverges from a source and strikes a flat mirror. Our focus questions here were: What will the ray-set look like *after* reflection? And where will the reflected set of rays *appear* to be diverging from? We used the law of refection to construct the reflected set. The changed ray directions gave rise to a new (re-directed) set of rays. This transformed set appeared to be coming from someplace else, and this led us to invent the concept of *image*.

B.

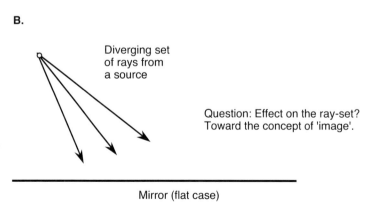

Diverging set
of rays from
a source

Question: Effect on the ray-set?
Toward the concept of 'image'.

Mirror (flat case)

Manipulating light, the basis of optical instruments

A useful way of looking at the operation of mirrors is that by using reflection we can effectively *redirect* a set of light rays coming from an object, in a way determined by the law of reflection. We will see that this 'transformation by reflection' of rays sets into other ray sets gives rise to the phenomenon an *'image'* of the original object. An image is generally in a different location than the original object. This possibility of *controlling* light with suitable mirrors to produce images turns out to be very useful, allowing us to devise *optical instruments*. The mirror is in fact an optical instrument! Another is the periscope, which uses two flat mirrors to let us 'see around corners'.

The mirror is about the simplest kind of optical instrument there is. You might not have thought of a mirror as an optical instrument, because it is so simple and familiar, but it acts to change the direction of light, and this has useful optical effects

Revisiting the prelude demonstrations

Now that we have explored the phenomena of reflection and developed the concept of image, it is time to revisit the demonstrations in the prelude! Look at them again, now that you have developed knowledge to understand them, but look with a scientist's eye this time.

> *"We shall not cease from exploration, and the end of all our exploring will be to arrive where we started, and know the place for the first time".* T. S. Eliot.

This time think about the demonstrations and phenomena in terms of the underlying science, and explain what is going on.

Notes

Flat and curved mirrors: looking ahead

A natural next question to ask is: what if the mirror surface were **curved** rather than flat? How would this affect the diverging ray set on reflection? How would this affect what one would 'see' in a mirror? You might suspect that changing the mirror geometry would change the ray-set geometry, and hence the images formed – and you would be right!

Let us quickly compare the situations for reflection from different types of mirror Reflection of ray-sets can be by either flat (plane) or curved mirrors. Curved mirrors can be either convex or concave. (There are demonstration mirrors in the lab – take a look at each kind.)

The figures below show diverging incident rays approaching three kinds of mirror surfaces: flat, convex and concave.

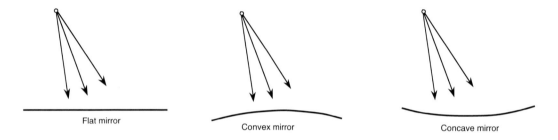

Flat mirror   Convex mirror   Concave mirror

The three situations differ only in geometry of the surface. We will want to explore how this affects the ray-set on reflection, and compare effects for the three cases.

And that is the topic of the next chapter! The basic physics principles and procedures will be the same in each case of course, but the geometry will differ, and that makes for interesting new effects.

These effects are the basis of other optical instruments, which use *curved* mirrors to manipulate light. In this case, not only is the image in a different place than the object, but it can be larger, smaller or the same size. For example the reflection telescope uses curved mirrors to produce magnified images of distant objects.

Curved mirrors and images are discussed in the next chapter. We will treat curved mirrors not as a 'separate topic' but as another example of the reflection of a ray-set. Thus ray tracing for curved geometry will be an 'extension of knowledge' for us.

Note

While we are talking about manipulation of light and optical instruments, let us mention that another phenomenon which can be used to change the direction of light is *refraction*. This is dealt with in later chapters, and is the basis of yet another class of optical instruments.

David Schuster 091115

218

PROBLEMS

REFLECTION, MIRRORS AND IMAGES

PROBLEM CATEGORIES

A. BASIC REFLECTION PROBLEMS
B. PROBLEMS ON IMAGES – PLANE MIRROR CASE
C. CALCULATION PROBLEMS ON PLANE MIRROR REFLECTION
D. COMBO QUESTIONS ON REFLECTION

A.   BASIC REFLECTION PROBLEMS

1. Sun's rays reflected from a puddle

Clouds are blocking the sun's rays except for a gap, and a narrow beam goes through to hit a puddle of water.

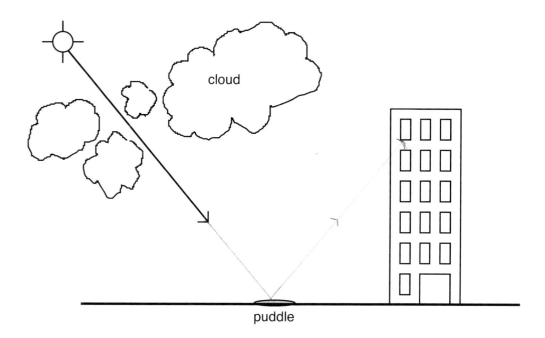

cloud

puddle

a) The beam reflects off the puddle and strikes a tall building. At which floor of the building will the beam arrive? Show your construction. 6th

b) What will people at the window at this floor see when they look at the puddle? a bright spot of light

c) What will people on higher and lower floors see reflected in the puddle? the reflection of the sun

(Question idea thanks to Group 6, Tue 11 am, 030211)

2. Bright area after flashlight reflection

A flashlight produces a diverging beam of light as shown, directed at a mirror on the wall.

ROOM

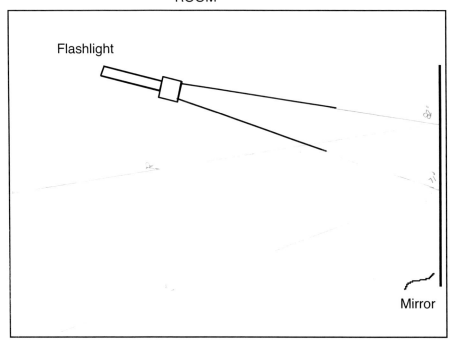

Flashlight

Mirror

Do accurate ray construction to show the size and location of the bright spot produced on the opposite wall.

This is due to the angle of the flashlight and the angles of the light reflecting off the mirror

3. Double reflection at corner mirrors

A room has two mirrors on adjacent walls as shown (top view of room). A laser pointer shines a narrow pencil of light toward one mirror as shown.

ROOM

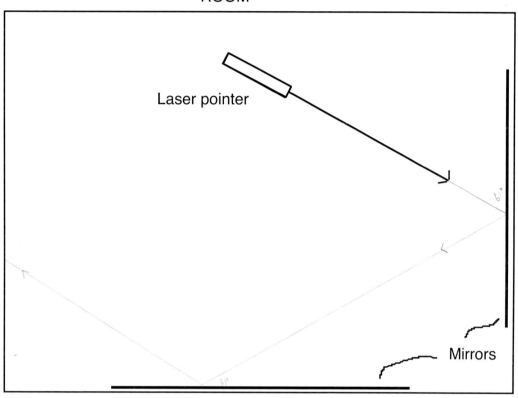

a) On what wall will a spot of light occur, and where exactly? Do an accurate ray construction, and mark the spot on the wall.

b) After the double reflection, how does the *direction* of travel of the light beam compare with its original direction? By inspection, does it seem to be parallel? Or not parallel? Same direction? Opposite direction?

c) *Prove* that after two reflections by mirrors at right angles, the light ray will be parallel to the original ray. This will need some geometry.

Note: This setup of two mirrors at right angles is called a 'corner mirror'. A corner mirror has the interesting property of exactly reversing the direction of an incoming light beam – no matter what direction it comes in at. Check this if you like: try another ray coming in at a different angle, and see what happens to it.

4. Mirror in a black box

An instructor puts a big black box, open at the top, on the bench at the front of the classroom. The class watches as the instructor lowers a mirror into the box and then shines a laser beam into the box at a certain angle (ray A). To everyone's surprise, a ray comes out of the box at a different angle (ray B).

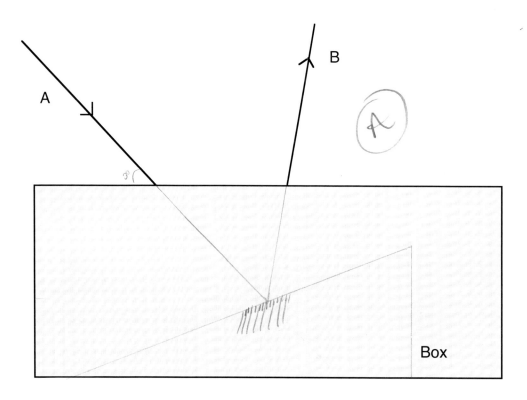

Students offer various ideas:

i.   That the mirror is a special kind that does not obey the usual law of reflection.

ii. That it is an ordinary mirror, but placed at an angle in the box.

a) Before announcing the discovery of a new kind of mirror, check if option (ii) could be true. That is, try to find a location and orientation of an ordinary mirror that would give this result. Show your constructions, explaining what you are doing. Draw what might have been in the box when the instructor lowered the mirror into it.

b) Suggest a testing experiment to test which of the two explanations might be best, by working with the consequences of each.

B. PROBLEMS ON IMAGES – PLANE MIRROR CASE

5. Reflection of diverging rays from a mirror

Three (of many) rays are shown diverging from a point light source and striking a mirror.

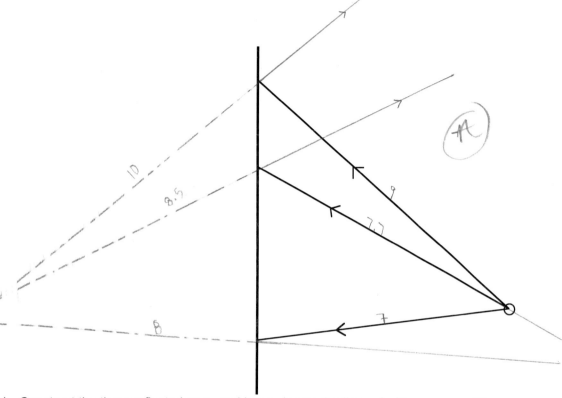

a) Construct the three reflected rays, and hence locate the 'image' of the source. (4)

b) How does the distance of the source from the mirror compare with the distance of your constructed image from the mirror? Compare by actual measurement on your diagram. (2)

6. Status: primary or secondary principles?

Consider the principle that "the image in a plane mirror is the *same distance* behind the mirror as the object is in front". This principle is ….

    i. a primary (fundamental) principle in its own right.

    ii. a secondary principle, in that it is a consequence of a primary principle and can be derived from it.

Support your answer.

Similarly consider the principle that the *size* of the image in a plane mirror is the same as the object size. Is this a primary or secondary principle in physics?

Not sure

You have to add in human error

7. Reflection of street light from a wet road

There is a pool of water near a streetlight one rainy night. Rays are shown diverging from the light and going toward the water, where they will reflect from the water surface.

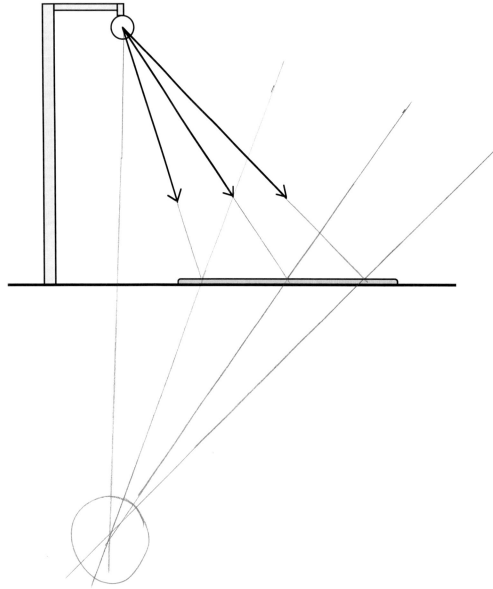

a) Extend the rays till they hit the water surface, and then accurately *construct* the three reflected rays. State what powerful principle you used in doing so.

b) If you are standing somewhere to the right in the picture, looking toward the water surface, and these reflected rays came to your eyes, where would it **seem** that the rays are diverging from? Show this location on the diagram.

c) An **image** of the lamp would be seen to be:
   A. Above the water surface
   B. At the water surface
   C. Below the water surface.

d) At what *distance* above or below the water surface would the image of the lamp seem to be?

8. Image in a mirror

A small light bulb is placed near a flat mirror as shown.

An observer Diane looks toward the mirror. The diagram shows Diane's two eyes, and a ray construction showing how rays enter them after being reflected from the mirror.

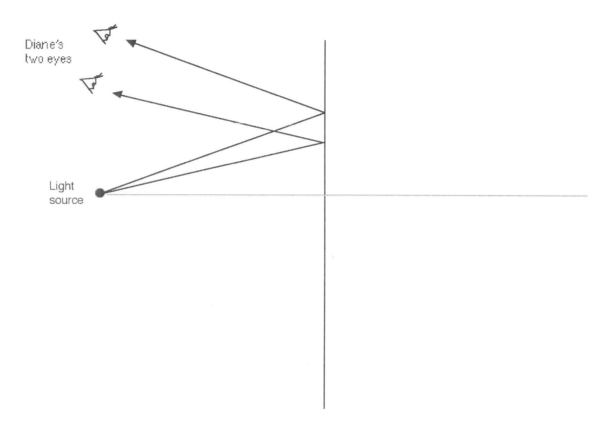

Diane's two eyes

Light source

Jim looks from somewhere around here

a) Do a construction to locate the 'image' that Diane sees.

b) Another student Jim is in a *different place*, i.e. somewhere in the area shown. If he looks toward the mirror, what would he see?

   A. He would not see an image at all.
   B. He would see an image but in a different location from that shown above
   C. He would see an image in the same location as that above.

   Support your answer with a careful ray diagram. Show Jim's **two** eyes on your diagram.

c) If a sheet of paper were placed at the location of the image seen by Jim in (a), then which one of the following would be true?

   A. There would be one bright point of light on the paper.
   B. There would be more than one bright point of light on the paper.
   C. There would be a fairly even illumination of the paper
   D. There would be no light on the paper.

   Explain your reasoning.

d) A student asks you: "OK, Jim sees an image behind the mirror, as if there is a light source there. So if I actually go behind the mirror and look at that spot, will I see it there?" Ask her a question or two, to lead her to the answer.

9. Ceiling lamp seen in a shiny floor while walking

The floors in Wood Hall corridors are pretty shiny, and you can see the ceiling lamps reflected in them. Try this: look at the image of a ceiling lamp in the floor ahead of you. Then start walking forward: what happens to the lamp image you see? Explain why it moves forward as you walk, and explain why it moves slower than you do! How much slower, exactly? What is the situation when you catch up with the image? And what happens next if you keep walking?

(You could also try this at home if you have a shiny floor somewhere, or in a public building etc)

(Discovered 030309.1900 while walking in Wood Hall, having seen it every day for a year previously without really seeing it!).

C. CALCULATION PROBLEMS ON PLANE MIRROR REFLECTION

10. Bob and Ann at the office - calculation

Bob and Ann work in an open-plan office. The boss puts in a partition to block them from having a direct view of each other, as shown in the office plan (view from the top).

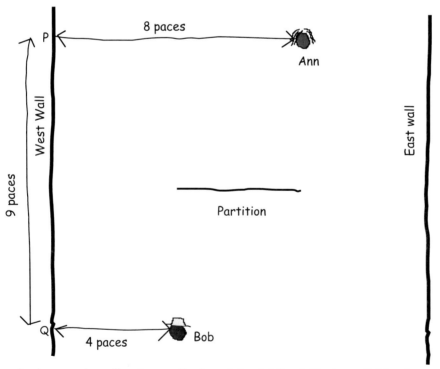

Bob has a brainwave: he will put a small mirror (about 1 ft x 1 ft) at a suitable place on the West wall so that he can see an image of Ann from his chair. He decides to use his physics and math knowledge to *calculate* where to put the mirror. He starts by making a sketch, and measuring distances by pacing them off, as shown on the diagram.

a) Where along the wall should the mirror be placed in order for Bob to see Ann? That is, how many paces from the spot Q? Do this problem by **mathematical calculation.**

(Suggested procedure: on the rough sketch above, draw in approximate light rays *freehand*, just to show the geometry of the desired situation. Then apply ideas of triangles and ratios to find the required distance along the wall. Annotate your working. Do NOT do this problem by scale construction).

b) Notice that we have deliberately given you a rough freehand sketch for this problem, rather than an accurate scale diagram. Why? That is, if a problem to be worked using mathematics, do you think it is more fitting to give a sketch, or a scale diagram? Discuss briefly.

11. Using a mirror and math to find the height of a tree

A student Debbie wants to use physics to find the height of a tall tree. Due to heavy cloud cover she cannot use the tree's shadow, so she invents another method. She lays a mirror on level ground some distance from the tree, then walks backward while looking into the mirror, until she reaches a spot where she sees the top of the tree in the mirror.

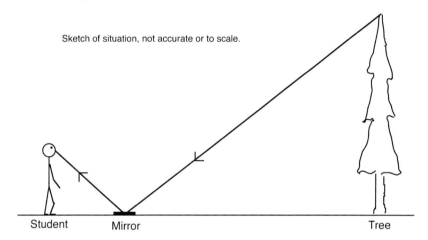

Sketch of situation, not accurate or to scale.

Student   Mirror   Tree

In this setup, she finds that she is 3 meters from the mirror, and the mirror is 15 meters from the tree. Her eyes are 1.5 meters above the ground. She draws the rough sketch above, not trying to be accurate, nor to scale.

a) Use *mathematical* methods to calculate the height of the tree. (Label distances with symbols and values. Proceed mathematically using symbols first and substitute numbers at or near the end. Justify any equation you write).
   10

b) If Debbie moved a bit further back, away from the mirror, what would she see in it?
A). Nothing. B). The top of the tree. C). The cloudy sky. D). A lower part of the tree.   2

c) For this method to work it is important that the mirror be exactly horizontal. (Why?) When Debbie first places the mirror on the ground, can you think of any clever (optical) way she can make sure it is horizontal, e.g. by looking down into it and using reflection ideas, or some other optical way?   3

D. COMBO QUESTIONS ON REFLECTION

12. Bob and Ann at the office

Bob and Ann work in an open-plan office. The boss puts in a partition to block them from having a direct view of each other, as shown in the office plan (view from the top).

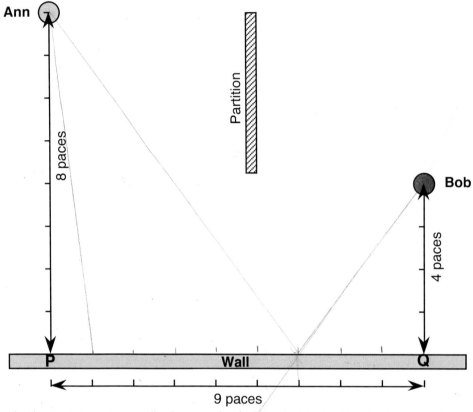

Bob has a brainwave; to put a small mirror on the wall so that by looking in the mirror he can see Ann from his chair. The problem is: where on the wall to place the mirror? He decides to use his physics and/or math knowledge to work it out, in three different ways, as follows:

a) *By construction from basics – using the reflection law and trial ray tracing.* Trace the paths of some light rays from Ann, reflect them and find one that goes on to Bob, and hence mark the right spot to put the mirror.

b) *By construction using the secondary idea of image.* Mark where an image of Ann would occur if there were a wall mirror, and use this to find where the small mirror must be placed. Is this method easier or harder to use than (a)?

c) *By math calculation.* Bob then uses physics and math knowledge to *calculate* where to put the mirror. He paces off distances in the office, and marks the values on a sketch of the situation, as in the diagram above. Use the math of triangles and ratios to calculate where along the wall the small mirror should be placed.

d) To Bob, where does the *image* of Ann appear to be located?

      [ ] In front of the mirror.   [ ] On the mirror surface.   [ ] Behind the mirror.

e) Will *Ann* also be able to see *Bob* in the mirror, or not? Explain.

f) The boss finds the mirror and removes it in the interests of getting work done. Nothing daunted, Bill decides to put a mirror on the *East* wall instead. Can he put it exactly opposite the spot on the West wall where he put the first mirror, or won't that work? Explain with the aid of rough rays sketched quickly on the figure above.

Note: This mirror problem looks very different from shadow problems or aperture problems. Yet can you see that the underlying mathematics used is much the same? There is a certain unity in all these situations, because they are all consequences of straight line light travel, and in all cases triangles are involved. So once you understand the underlying approach, it can serve you well in various situations!

13. Who can you see in the mirror?

The figure shows atop view of nine people, including yourself, seated at a long banquet table, with a mirror opposite. Your seating place is shown black.

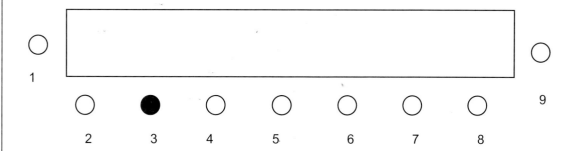

a) Of your fellow diners, who would you be able to see in the mirror?

Do this problem in two ways.

i. By accurate construction of some appropriate light rays (as if you could 'ride light rays and obey the law'). Do not use the concept of image.

ii. By accurate construction again, but this time using the concept of 'image', and using what we know about image location for a plane mirror.

b) Which of your fellow diners would be able to see YOU in the mirror?

14. Where to aim?

A laser pointer is located at A. You wish to aim the pointer at the mirror in such a way that the reflected beam hits an object at B.

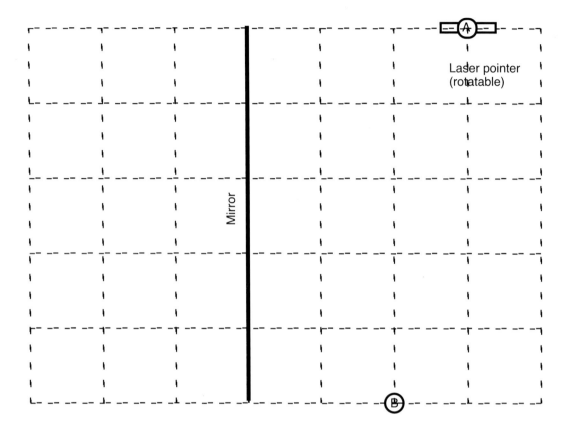

d) In what direction should you aim the pointer? Show the direction on the diagram.

e) Are there a number of ways you can tackle this problem on paper? Which seems easiest? Why?

f) Suppose this was a real life situation, and you were standing at point A with the laser pointer, looking toward the mirror. Practically, how would you know where to aim to hit B?

Chapter 10

CURVED MIRRORS: REFLECTION AND IMAGE FORMATION

In this chapter we explore the reflection of rays and by curved mirrors and the production of images The difference between this and the previous case of plane mirrors is simply the shape of the mirror. Thus we expect that most if not all of the physics ideas and procedures of the previous chapter will carry over to this one, but we are curious how reflection and images will be affected by the different mirror shape. Note that curved mirrors can be either convex or concave, and may have various degrees of curvature; and we also wonder also about the effects of these aspects.

SECTIONS

PROBLEMS

10-1 PRELUDE

Demonstrations and visuals involving reflection by curved mirrors

The instructor may introduce some lead-in demonstrations and visuals, to arouse interest and curiosity, as a prelude to our exploration of curved mirrors. We have various mirrors available: convex and concave mirrors of various sizes and curvatures, a curved shaving/cosmetic mirror, a vehicle side-mirror, a store security mirror, shiny soup spoons, a 'floating pig' demonstration, and more.

We will not try to *explain* these demonstrations immediately, because we first need to turn our analytic and creative energies to understanding reflection and image formation at curved surfaces. Then we will come back to these demonstrations, armed with the knowledge to explain them.

10-2 REFLECTION AND IMAGE FORMATION BY CURVED MIRRORS – Introduction

Mirror surfaces can be either plane, convex or concave. All three shapes are drawn alongside. A light source is shown to the left of each.

In the previous chapter we treated reflection and image formation by a *plane* mirror. But what If the mirror is *curved*? Interesting questions arise:

> *How will light rays be reflected? Will images be formed? If so where will they be located and what will they be like? What differences will there be from the plane mirror case? Does the degree of curvature of the mirror affect things?*

We will treat reflection from curved mirrors as a *problem* to be tackled – an extension-of-knowledge challenge! We view it that way – as an application of our existing knowledge to a new situation – because we suspect that there will be no new fundamental principles involved, just the basic law of reflection applied to a curved geometry.

Of course we anticipate that some new and interesting effects might arise from the curved geometry! You have seen some of those in the prelude demonstrations.

Note that conventional textbook chapters on curved mirrors usually *give* you the facts, derivations and conclusions. That is, they do the thinking for the learner and *present* everything – often with much mathematics – to be absorbed and understood

By contrast, in our inquiry approach, we will take it as a challenge to work out for ourselves what happens if the mirror is curved, starting from basics, since we know the underlying powerful ideas. It will also be a good test of our real understanding of the basics and how to use them!

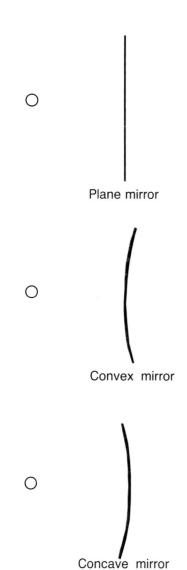

Plane mirror

Convex mirror

Concave mirror

10-3 RAY REFLECTION AT CURVED SURFACES

How does light reflect from a curved surface?

Previously we explored and understood the reflection of light at a plane surface. Now we ask:

> What if the surface is curved? How do we apply the law of reflection to a curved surface?

Think about the situation for yourself, and how to address this. A curved mirror surface is shown below.

Ideas and discussion

The arc below represents the mirror surface. Draw in a few rays coming toward it. Think about how they might reflect. Can you suggest of a way to apply the basic law of reflection where they strike this curved surface?

For this purpose, how can we think about the surface just near the point where the ray strikes? What procedure might you use in constructing reflected rays?

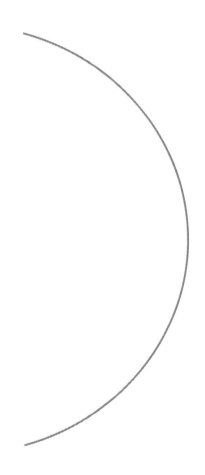

Constructing rays reflected from curved surfaces

Now try reflecting rays from both convex and concave surfaces, using accurate construction. Consider various methods of making equal angles and/or drawing tangents to the curve. The instructor will guide. Use large diagrams.

You will find there are some distinct cases:

a. **Convex mirror**

b. **Concave mirror cases:**

   i. Source close to mirror,

   ii. Source far from mirror,

   iii. Source intermediate,

   iv. Source at center of curvature.

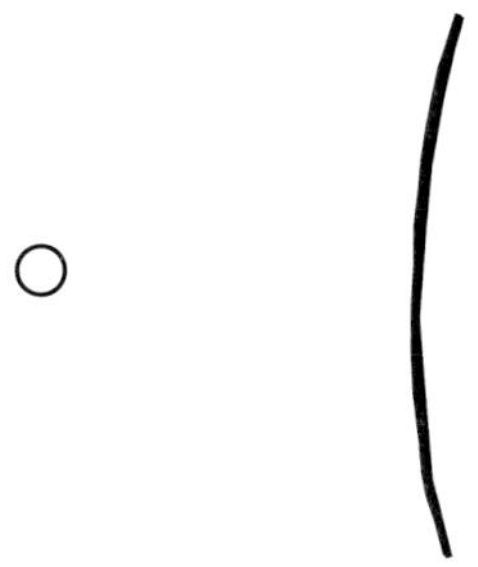

Convex mirror

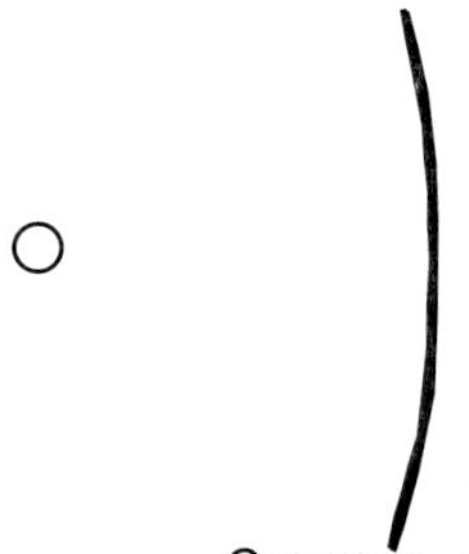

Note that 'close' or 'far' mean close or far in relation to the radius of curvature of the mirror.

Concave mirror

The instructor can set you some large-scale diagrams to practice the procedure on.

10-4 *IMAGES* DUE TO CURVED MIRRORS

Question: Are images formed by curved mirrors, and if so how do we construct them?

A challenge task

Simple case first: a single point source and a **curved** mirror. how can we find the image formed by light reflection from the curved surface?

Well, we have the same procedure available as for the plane mirror, that is: consider a diverging ray-set from the source point, reflect the rays at the surface (curved this time), find what happens to them, and see if the reflected set appears to come from a particular place.

Can you do this, using just the law of reflection and the geometry of the surface?

Reflection of diverging ray-sets by curved mirrors

Our aim then is to find out what happens to a diverging ray-set ('DivRaySet) from a point source when it encounters a curved mirror. This will tells us if an image is produced.

On the right we repeat an earlier diagram of diverging ray sets approaching three different shapes of mirror. We know the basic physics involved in all of these – any ray incident on any mirror surface is reflected off at an equal angle. Thus in principle we can simply apply this law to the rays in the set, whatever the shape of the surface.

But how do we apply the law at a *curved* surface? That is, how do we construct a reflected ray at an equal angle to the incident ray? We worked this out in the previous section!

The physics of reflection will be the same whatever the shape of mirror, so the basic procedure for ray tracing and reflection will be the same for all mirrors. Now we have the task of working out what difference the curved geometry makes, for possible image formation.

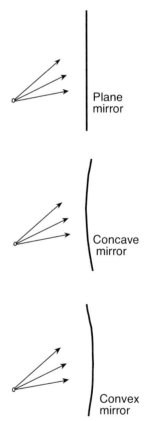

Plane mirror

Concave mirror

Convex mirror

1. CONVEX MIRRORS - IMAGE FORMATION

We start by studying a **convex** mirror.

A. Exploring the real thing: effects from a convex mirror

First, let's explore what effect we get from reflection in a convex mirror. Use a small light bulb as source, place it near the mirror, and observe the image produced by reflection. Note that is very useful to place a convex and plane mirror side by side, so the effects can be compared.

What do you observe about the image in the convex mirror? Where do you think it is located? How do the images in the two mirrors compare? And their locations? You can also try varying the distance of the source from the mirrors.

Report on all this.

Convex mirror

Plane mirror

Note

In practice if you do not have a very small 'point' source, but one that has some size, you will probably also notice the *size* of the image, not just its location. Image size is not our concern at the moment, since we are starting with point sources for simplicity, but note that curved mirrors do produce interesting size effects, which we can deal with later.

B. Theoretical treatment using the ray model and the law of reflection

Let us work out by ray construction what happens when light from a point source is reflected by a convex mirror, and locate any image formed.

To represent the convex mirror surface, draw a large arc on a whiteboard or sheet of paper (see next page.) Choose a location in front of the mirror for the point source, and mark its position.

The aim is to find what happens to a set of diverging rays from the source after reflection from the curved surface, and hence find the spot where they *seem* to be coming from. In this way we will locate the 'image' formed by the convex mirror.

*Note: Reflecting rays from a curved surface will need some care! You have to construct equal angles in a curved situation, as accurately as you can..

Visualizing the basic physics in operation: Ride a ray and obey the law!

Remember that the basic physics is the important thing, i.e. how a ray travels and reflects. For teaching young students there is a useful way to visualize this that 'puts them in the picture'. You can think of yourself as "riding a light ray" in a straight line toward the mirror, and then "obeying the law" at the mirror, and then continuing riding straight in the reflected direction. Make a cartoon sketch of a stick figure riding a light ray.

Riding a ray: sketch

Ray construction for convex mirror: next page

Locating an image by ray construction – convex mirror

Make a larger diagram on a whiteboard, to the same proportions, for a group presentation

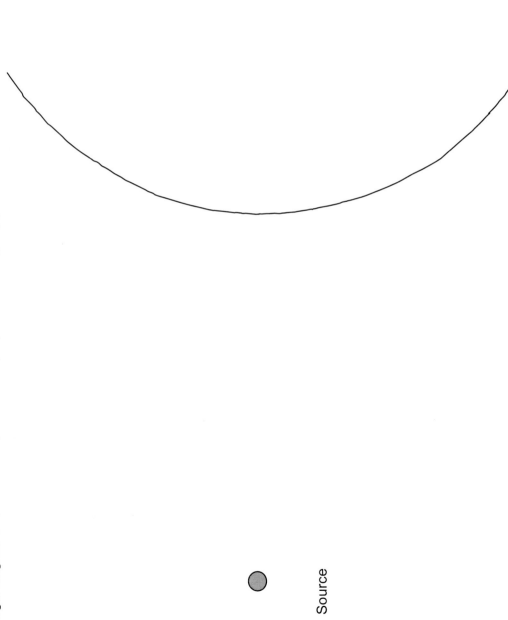

Source

Image distance

When you are done with the constructions, note where the image is located, compared to the previous situation of a plane mirror. Does the previous 'equal distance' rule still apply, or not? If not, in what way does the convex situation differ?

How do image locations compare for the plane and convex cases?

Can you see conceptually **why** this should be the case, from comparison of ray diagrams and what is happening to the ray-set at the two shapes of surface?

C. Comparing experiment and theory

Compare the image location you observed in the lab with the location you found theoretically using ray construction.

Are experiment and theory consistent?

THE *FEATURES* OF THE CONVEX MIRROR CASE

Look at the ray diagram for image formation in the convex mirror case. Try to identify the main features, that would characterize this case, and list them alongside.

Then check your ideas against those offered below:

Features: For a convex mirror ...

1. The image is behind the mirror, and virtual - the reflected rays only *appear* to be diverging from that spot.

2. The rays striking the mirror are reflected strongly 'outward' due to the convex shape of mirror – hence the set diverges more after reflection than before reflection. (In the plane case the degree of divergence was the same before and after).

3. This causes the virtual image location to be *closer* to the mirror than the source is. (Note that the *explanation* for this is point 2 above and is evident on the diagram)

4. Thus the previous 'equal distance' rule we found for the plane mirror case does **not** apply to the convex case! Instead, we may state a qualitative rule for image location in a convex mirror:

Effect of varying the source distance
– how does the image behave?

So far we have done an image construction for just one particular source distance from the curved mirror. What if we *vary* the source distance? That is, what will happen to the image if we move the source closer to the mirror or further from the mirror?

Here we will do the same kind of conceptual thinking, using our ray model, as we did earlier for the plane mirror 'variation' case.

We could of course draw new constructions for a number of different source distance, but maybe we can 'see' what the behavior will be by looking at the ray diagram and *imagining how it would change* if the source distance changes.

Try this for the convex mirror case now! You can do it in your imagination, or use sketches on paper.

What do you conclude?

Note: more than one way to do it

Note that there are at least two ways to do the above. Either keep the DivRaySet the same (i.e. keep the same divergence angle), or keep the place on the mirror the same where the rays strike (which means choosing a differently diverging ray set). The instructor can discuss. Which method did you use? If you like, try the other method also, and see how it gives the same result (as it must).

Idea power!

The nice thing about thinking about things in this conceptual way is that afterwards you will understand **why** the image behaves as it does as you vary the source distance. That is, you will have a conceptual grasp of what's going on, will be able to predict the effects, and will be able to explain them to others. And you should be able to do this anytime in the future, long after you have forgotten what actually happens.

Check your theoretical predictions against reality

You have predicted what will happen to the image by using the ray model; that is, this is a theoretical prediction. Of course you should also try the real thing: e.g. move a source closer to a convex mirror and see what happens to the image location – does it move closer, stay put, or move further back? Is this in accordance with your theoretical prediction?

Effect of degree of curvature

Note that mirrors also vary in how curved they are. The surface can range from only slightly curved to strongly curved. We wonder whether this might also make a difference to what happens, and if so, what kind of difference.

The instructor may lead a discussion of this issue, and how to explore it both theoretically and practically. Your current knowledge is enough to understand it for yourself, with careful thought!

Note on comparative or relative distances

What is really involved here is not just the radius of curvature of the mirror, but the radius **in comparison to** the distance of the source from the mirror. Thus many of the previous results on distance variation can be re-interpreted when we turn to curvature variation. The instructor can discuss this with you.

2. CONCAVE MIRRORS - Image formation

The physics will be the same for all shapes of mirror, convex, concave, and plane, but the geometry is different, and we image that this might make a difference for the location and nature of images formed. Even without doing it in detail, we might expect in advance that there should be a difference for a concave compared to a convex mirror, just because the surfaces curve opposite ways, and this will surely affect the directions of reflected rays. Let us explore this, both practically and theoretically.

Concave mirror

A. Exploring in practice - effects from a concave mirror

Let us first explore the effects in practice, with real mirrors. Again, it is useful to do this *comparatively*, by setting up convex, plane and concave mirrors side by side, and observing and comparing the images that arise in all of them, for a given source location.

Set up the trio of mirrors, and use a small light bulb as source. Start with the source very close to the mirrors, and gradually move it further and further away, noting the effects in each mirror. Note both the appearance and location or the images. Note that you will sometimes find it hard to judge location, and sometimes it may be unexpected!

Call on the instructor to guide you. It is best that you as the observer remain quite far from the mirrors. The instructor will also help you with the idea of a 'real' image in some of the cases, and show how to 'capture' it on a screen.

Record your observations and insights.

B. Graphical 'simulation' treatment of concave mirror image formation – using accurate ray tracing and the law of reflection

The best way to appreciate what is happening (and why) with concave mirrors is to **simulate** the various cases by accurate ray reflection of ray sets. First a set (or at least two) diverging from the source, then being reflected from the concave mirror, and then proceeding again as a set with changed direction.

Do the ray simulations by drawing accurate diagrams for the following cases: (Important!)

1. Point source position close to the concave mirror (closer than half the radius of curvature).
2. Point source far from the concave mirror (further than the radius of curvature)
3. Point source at an intermediate distance.

Do large constructions, on three separate sheets, and on class whiteboards. Report and compare your results.

C. Image distances

After constructing rays and finding image locations, do you agree that the image-location rule for plane mirrors does NOT apply to concave mirrors? It is much more complicated for concave mirrors, and we will not go into the mathematics, but just be content with being able to locate images for any situation, but accurate ray construction, i.e. by simulation the behavior of light itself.

Note how we get very different results for image location when we apply the same physical law to different geometrical situations

.

D. Comparing theory and experiment

Are your theoretical results consistent with your experimental observations?

10-5 IMAGES OF *E X T E N D E D* SOURCES IN CURVED MIRRORS

Challenge: to construct the image of an **extended** source in a curved mirror

This is a 'physics & geometry' challenge! To construct the image of an extended source in a curved mirror, by application of the basics.

You know all the physics required to do this, so you should be able work it all out from first principles! But it will still be a challenge to put it all together for a source with size and shape.

Use an arrow with head and tail as the extended source. (Why?) Note that there will be various cases to consider: either convex or concave mirrors, and various distances of source from mirror.

i. We will start with a convex mirror.

Given: a curved mirror arc drawn on large paper, and an extended object (arrow) drawn near it. Start with the case of a convex mirror.

Paper provided. Your group can also do it on a large whiteboard: the instructor will check that you have a practical arc size and curvature and suitable y source location .

Treat this a group task: trace rays, and construct the image. When you are done, look at both the location and the size of the image, compared to the object.

What is the effect of a convex mirror on image size?

Note:

If you managed to do this successfully, you have worked out something from basic concepts and principles, that is usually treated only in more advanced texts. Congratulations! In those texts it is usually done rather mathematically, with less conceptual understanding.

ii. Now turn to concave mirror cases.

You know that the location and nature of the image depends on distance of the source from the mirror, so there will be different cases to construct.

244

Looking at your *own* image in plane and curved mirrors

Thus far in investigating mirrors and images, you deliberately used some object, other than yourself, as the source of light. You acted simply as the observer.

However our most common experience of mirrors is when we see ourselves in them. In this case, we have the dual roles of both object and observer! It is interesting to see how we appear in plane, convex and concave mirrors. It helps if you place the different mirrors side by side, so that the comparison is obvious.

Report on what you observe.

You can also see the effect of distance − start from being very close to the three mirrors and gradually move further and further away, noting what you see.

David Schuster 090818

PROBLEMS

Curved mirror reflection and image formation

A. REFLECTION FROM CURVED MIRRORS

Ray tracing problems involve accurately reflecting rays which encounter a curved surface, either convex or concave. Just the basic law of reflection is involved, for a curved surface geometry. You will need to construct equal angles at a curved surface. The credo, as before, will be: *"ride a light ray and obey the law"*, i.e. obey the reflection law when you come to a surface.

Note that there are various example and assigned problems in the chapter. Further problems will be provided by the instructor.

B. IMAGE FORMATION BY CURVED MIRRORS

Image in a concave mirror

A concave mirror is placed on a table, and a small (maglite) light bulb is clamped above it, as shown. The task is to locate the image produced, by constructing rays as accurately as you can. Explain the method you will use. Do the ray construction. Also show roughly where your eyes must be to see this image.

Chapter 11

REFRACTION

– light going from one medium to another

We found earlier that light can travel not just in air but also in various other transparent media, such as vacuum, glass, water and plastic. In each of these, light travels in straight lines.

The interesting question arises – what happens as light goes from one transparent medium to another? Does it just keep going in the same straight line, as if nothing had changed, or does something happen to it at the interface, and if so what?

It turns out that the answer has dramatic practical applications and real-life consequences! If you wear glasses or contact lenses your improved vision depends on what happens to light between one medium and another. Optical instruments such as cameras, magnifiers, telescopes, microscopes, projectors, etc are have lenses based on the effect, and the operation of devices like CD and DVD players makes use of light and lenses. Natural phenomena such as the rainbow arise when light goes from air into raindrops and out again.

Thus it is important to explore the phenomenon scientifically, try to understand what is going on, and discover the laws of nature that apply. So we will first explore the *basic phenomenon* qualitatively, to see what happens when light goes from one medium to another. Then we will extend our quest to see if *images* can perhaps be formed using the effect. Then we will try to devise a way to trace rays accurately between media, empirically. Finally, in the next chapter, we will do a *quantitative* investigation, seeking a *law* for the phenomenon.

SECTIONS

PROBLEMS

11-1 PRELUDE
Advance demonstrations and visuals

The instructor may introduce some lead-in demonstrations and visuals, to arouse interest and curiosity, as a prelude to our own systematic investigations. We have various pieces of equipment, e.g. light beams, lasers, blocks of glass, prisms, square plastic rods, water in containers, a fish tank, cups to pour water into, lenses (e.g. your spectacles or a magnifying glass), and more.

We will not try to *explain* all these demonstrations in this prelude, because we need to learn more about the phenomenon before we can do that.

At the end we will return to this prelude and see the demonstrations with new eyes – and minds – and armed with the knowledge to explain them.

DEMONSTRATIONS

VISUALS

11-2 EXPLORING THE BASIC PHENOMENON

We begin with a *qualitative* treatment. First we explore the *basic phenomenon:* the behavior of light as it goes from one medium to another. Let us approach this as a mystery: as if we were scientists exploring what happens to light.

Focus question

> *What happens when a beam of light goes from one medium to another, e.g. from air into glass?*

We first explore this *qualitatively*, observing how light behaves as it crosses from one medium into another.

Qualitative exploring

The first stage is exploring, observing, trying things out, and seeing what happens. We start by observing what happens when a light beam goes from air to glass, and then explore the behavior as we vary the angle of the beam.

The setup is as shown. A narrow light beam, obtained using a slit in a ray-maker, is incident at some angle on a glass block, as shown.

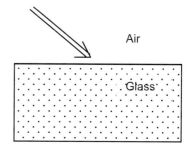

We are interested in what happens to the beam as it enters the block.

We also want to know the behavior as the incident angle changes.

First of all, simply "play with" the system! That is, set it up, try it out, see what happens, and vary things. Get a feel for the effect and how it behaves. Write a brief account.

NOTES

Naming the phenomenon – refraction !

Now that we have observed and explored the phenomenon, it is time to give it a name - scientists call it "**refraction**". Essentially, refraction is the **bending** of light as it crosses the interface between one medium and another.

Note that we did not name the phenomenon before observing it. Just as scientists do, we prefer to explore first and then give names.

Questions of what and why

Notice that while we explore the question of **what** happens to light at an interface, we do not tackle the difficult question of **why** light does this! The answer to that must be sought by finding out more about the intrinsic nature of light, and about the nature of the media through which it travels. For the moment, we must be satisfied, just like scientists of old, to simply describe the behavior as fully as we can, and try to discover a law that it obeys.

Notice that some light is *reflected* too?

Our main focus is on what happens to a ray as it passes between two media, e.g. air and glass. However you will have noticed something else happening too: not all the light goes into the glass, but some is *reflected*. You will also note that the reflected part seems to obey the law of reflection, so we already understand that. So be aware that some light is reflected, but now let's focus on our central issue, i.e. how the ray behaves as it goes from air into glass.

RECORD OF OBSERVATIONS – TRACING RAY BEHAVIOR

Now that we have a feel for things, let's make a good description by making a reasonably accurate record of the behavior.

Try a range of about four different angles for a light beam incident on a glass block, first 'head-on', then small, medium and large angles to the normal. Record your experimental observations by tracing incident and refracted rays on the paper beneath the glass block.

Then draw summary diagrams, and describe the behavior in words.

HOW THE EFFECT VARIES WITH ANGLE
– A previously-taken tracing of light rays

The figure below gives a previously-taken tracing, showing the refraction of light rays, for small, medium and large angles, all on the same diagram for comparison. Does it agree qualitatively with your own investigation?

On the left of the diagram, draw in one more case, for light incident normal to the surface.

REFRACTION BETWEEN AIR AND GLASS

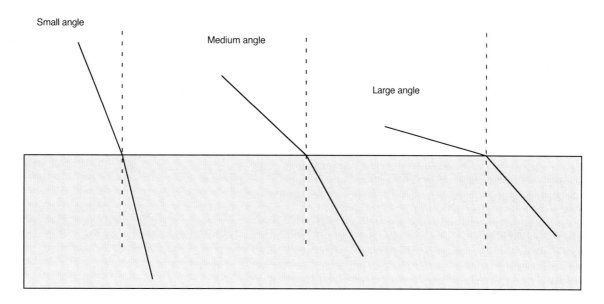

Compare the four cases, noting the behavior of the refracted ray as the incident angle increases. Then identify and write down the *features* of this behavior.

How does your feature list compare with this? At zero angle (normal incidence) there is no bending; at small angles the bending is small, but increases as the angle gets larger. At large angles (near grazing to the surface) there is a relatively large refracted angle, but that is clearly as far as things can go – that is the largest incident angle, so the corresponding refracted angle must be the largest possible!

253

The ray diagrams plus the paragraph above serve as a *qualitative* description of the refraction of light and how it varies with angle. Of course science would also like to have a *quantitative* description, where the angles are specified numerically, but that comes later. We will first work qualitatively, and find we can get quite far that way in our study of refraction.

WHAT ABOUT LIGHT GOING IN THE OTHER DIRECTION?

The question arises: what if the light beam travels the other way, i.e. goes from glass into air? Think about a way to explore this in practice. Ideas?

Then try it out and state what you conclude.

Does your qualitative refraction knowledge, summarized in the feature diagrams and listing above, serve to cover this 'reversed' situation as well?

11-3 REFRACTION IN DIFFERENT MEDIA

Focus questions

So far we have investigated refraction for light going between *air and glass.*

The interesting question arises: *what happens for light going between other transparent media?* For example, between air and water, or between air and plastic, or between water and glass?

And if refraction does occur for these media, how does it compare with the refraction we have already investigated, air-to-water? Does light bend the same, or different?

Investigation

Let's investigate! Take the case of light going from air to water.

You can use a semi-circular transparent "cheese box" as the water container, and send a narrow beam of light at the flat side. Then trace what happens to the ray once it enters the water.

Present your traced diagrams for small, medium and large angles of incidence.

Compare the refraction between air and water to that between air and glass. State your conclusions.

11-4 WHAT WILL OUR EYE SEE?
- if we receive a ray which has been refracted?

The 'as if' principle at work

If our eye views light which has been refracted, then it will be receiving light rays which have changed direction since leaving the source. Our eye however, will not 'know' this, it just receives whatever light comes in, and interprets this as a source in the direction from which the light arrives. Thus to the eye, it will seem "as if" the light source was in another direction than it really is.

At the moment we will just consider a *single* ray of light changing direction and entering one eye. (Later we will deal with images due to refraction, and consider two or more diverging rays).

Some demonstrations

The instructor will set up some demonstrations of how refraction affects what we see.

11-5 TOPIC KNOWLEDGE SUBASSEMBLIES
– for the basic refraction phenomenon

You can now devise knowledge subassemblies to sum up your knowledge of basic refraction.

Phenomenon diagram

Behavior (variation) diagram

The variation diagram can show how the effect varies with angle. (This can look similar to our earlier 4-part diagram showing separately the refraction variations, or you may choose to overlay the variations, though that may not be as clear).

Mental images

It useful to form 'assembled' mental images of these diagrams, to access whenever you think of basic refraction and its behavior.

Note

Later we go on to *Image formation* by refraction, which is more complex than the basic phenomenon, so we will likely want to make a 'principles and processes' diagram for that.

11-6 FUN APPLICATION: "PENNY IN THE CUP"

Activity: the mysterious reappearance of a penny

Put a penny at the bottom of a wide empty cup, toward the far side. Looking at the cup so that you can see the penny in it, move your head slowly downwards, until the rim of the cup obscures your view of the penny. Keeping your eyes in this position, have someone start filling the cup with water. What happens?

The idea is to observe the effect, then try to explain it. Illustrate with ray diagrams.

Explanation and ray diagrams

Classroom teaching and assessment video: viewing and analysis

We will show a video of a teacher reviewing refraction with her class, then using the penny-in-the-cup activity to assess her students' understanding.

Watch the video, analyze the teacher's approach to both review and assessment, note what the reaction of the students was to both, and write a commentary/critique. Did the students learn much, and did the assessment confirm this? You can also make suggestions for improvement, both major and minor. How do you think the teacher's approach to refraction compares with ours?

Then we will have a class discussion and the instructor will guide and give input.

11-7 FORMATION OF *IMAGES* DUE TO REFRACTION
– qualitative treatment

Image formation

Basically, we know that a ray of light bends at an interface. Because of the change of direction, if we view the ray with the eye, it will be seen as coming from a different direction than the source really is. Hence we will see an 'image' in a different direction than the actual source.

To locate an image more precisely, as far as both direction and distance are concerned, we need to trace at least two rays diverging from the source point, they change direction at the interface, and as a consequence appear to be diverging from another point – the image location.

Challenge – work it our for yourself

Take it as a challenge to figure out for yourself how images are formed due to refraction, and do (diverging ray constructions to locate images. At this stage you can draw qualitative diagrams; later we will do it more accurately, using either refraction templates or the (quantitative) law of refraction. You can do your ray diagrams on whiteboards or large sheets of paper. The instructor will lead you through this, as needed.

In working out how to locate images due to *refraction*, you might like to draw on the understanding that you developed earlier for the case of *reflection*.

The "as if" principle at work again. Notice how the "as if" principle plays a role here in determining where things appear to be located! The rays entering your eyes seem to be coming from a certain location – but the real object is not at that location! The rays from the object have been bent – and what you are seeing is an 'image' of the object in a different location.

Topic knowledge subassemblies

At the end, devise topic knowledge subassemblies for images due to refraction. These might include: a principles and processes diagram, a features diagram, a behavior (variation) diagram, and possibly a cases diagram for light going between media in one direction or the opposite.

Note

Thus we see that refraction, like reflection, produces *images*, because it changes the direction of light rays in a certain way. This 'manipulation' of light rays thus produces interesting and useful effects. If we deliberately control it, by choosing the media and the geometry of the surfaces, we can design devices for specific purposes. This is the basis of many optical instruments.

Example refraction problem (qualitative)

Spearing a fish from the river bank

A fish is under the water at the position shown, and a fisherman on the river bank hopes to spear the fish for dinner. One of the rays of light from the fish is shown going toward the surface.

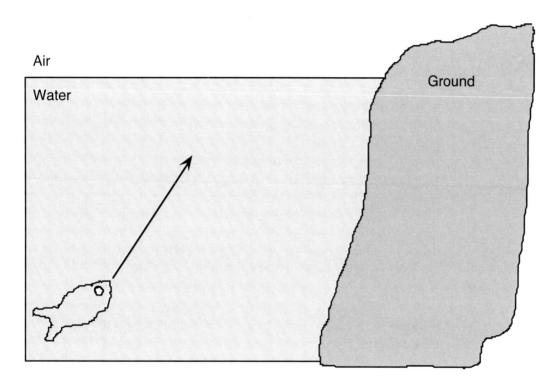

a) What happens to the ray as it passes through the interface between water and air? Draw in qualitatively the approximate path of the refracted ray which emerges into air.

b) Position the fisherman's eye so that she is able to see this particular refracted ray. That means she will see the fish as being along the line of this ray.

c) Based on just this one ray, will the fish appear to be higher or lower in the water than it really is? Explain. Show on the diagram the line along which where the fish will *appear* to be located.

d) If the fisherman throws the spear exactly in the direction that she sees the fish, what would happen? Would the spear *hit* the fish, or go *over* it, or *under* it?

e) So how should she aim if he wants to hit the fish? *At* the fish, *higher* than she sees it, or *lower* than she sees it? Note: this is a relevant question for spear fishing from the bank! And physics explains the technique that fisherman have learned from trial and error experience.

f) Suppose the fisherman is holding the spear pointing directly in the direction she sees the fish. What would the fish see? Would it appear *to the fish* that the spear was aimed for a direct strike or not? . Explain. So should the fish be afraid, very afraid?

g) So far we have made use of just one ray to reach some conclusions. With one ray we can find the direction of the image. If we want to find the exact location, i.e. both direction and distance, then we need to trace two or more rays. Do this, and hence locate the image of the fish. Because we are just using a qualitative treatment of refraction at the moment, the image location will be qualitative rather than precise. However that will be good enough to show the features of the situation and illustrate the procedure for finding images.

11-8 EMPIRICAL TREATMENT OF REFRACTION USING TEMPLATES

In this section we produce so-called 'refraction templates' which represent actual recorded directions of incident and refracted rays, in convenient diagrammatic form as a 'template' which can then be used to find the direction of a refracted ray if you are given an incident ray.

    A. Producing refraction templates

    B. Using refraction templates in problems

A. PRODUCING REFRACTION TEMPLATES

Producing a refraction template by ray tracing

Your *law* of refraction will enable you to *calculate* the direction of a refracted ray, for any incident direction. However there is another way of finding refracted ray directions: instead of calculating from a law, you can make a *refraction template,* which simply shows a convenient set of incident directions and provides the corresponding refracted directions. You make your own template from refraction measurements in the lab. The instructor will show you the idea. A template for air/glass refraction is shown in the figure.

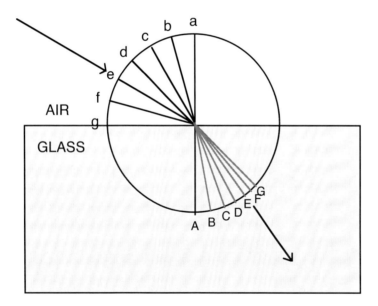

A set of rays on one side of the interface is labeled a, b, c, d, etc, while the corresponding set of refracted rays on the other side is labeled A, B, C, D, etc.

Thus the template contains refraction data of corresponding incident and refracted ray directions, across the whole range of angles, in a handy visual template you can carry with you!

Using the refraction template

You can use the template for tracing refracted rays in solving refraction problems. The figure shows a ray in air incident at the direction labeled 'e'. The template tells you that the corresponding refracted ray in glass will be in the direction labeled 'E'.

To use the template in refraction problems, you simply lay the template along the interface, note the direction of the incident ray relative to the labeled lines, and construct the refracted ray along the corresponding line.

The template only shows a limited set of ray directions, but for in-between directions you can *interpolate* reasonably well by eye.

Note that the template can equally well be used if the ray is incident from glass and refracts into air. In this case, if the ray is incident from direction E in glass, it refracts in direction e in air.

Making templates for air-glass and air-water refraction

The start of a large template is given on the next page. A set of labeled incident ray directions has been chosen at equal intervals. You produce the corresponding refracted ray directions by experiment using real rays, then draw them in on the lower half of the template, and label them A, B, C etc.

Note that in the first diagram we have only shown ray directions incident in the left quadrant, although naturally they can also be incident from the right. This is because the left and right situations will be symmetric, so you can just turn your template over if rays are incident from the right. If you prefer, you can produce templates that show both left and right quadrants, as in the second diagram.

Notes

You can think of a refraction template as a handy diagrammatic way of 'carrying around' your actual experimental refraction data! You could instead carry a tabulated set of data, but that is not nearly as cool, nor as easy to use.

Note also that you would also need a different refraction template for each set of media, for example one template for air/glass, others for air/water and air/plastic, and yet others for water/glass etc.

TEMPLATE 'STARTER' DIAGRAMS

1. Template starter showing left quadrant ray directions only. Right quadrant is symmetric to left. Complete the template by experiment, for use in refraction problems.

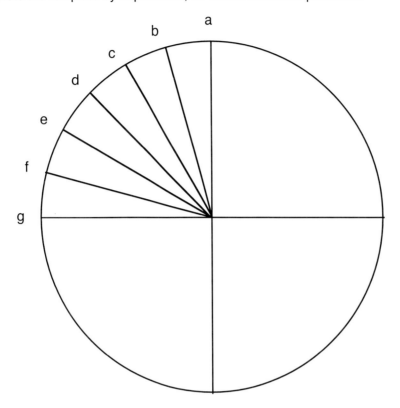

2. Template starter showing both left and right quadrant rays. Complete the template by experiment, for use in refraction problems.

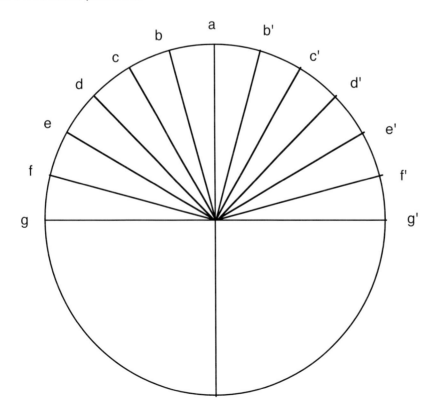

B. USING REFRACTION TEMPLATES IN PROBLEMS

The instructor will provide example problems, to solve by ray tracing using your own refraction templates.

11-9 POSTLUDE
– revisiting the demonstrations and visuals of the prelude

Now that we have explored the phenomena of reflection and developed the concept of image, it is time to revisit the demonstrations and visuals in the prelude! Look at them again, now that you have developed the knowledge to understand them, but look with a scientist's eye this time.

Think about the demonstrations in terms of the underlying science, and explain what is going on.

> *"We shall not cease from exploration, and the end of all our exploring will be to arrive where we started, and know the place for the first time".* T. S. Eliot.

PROBLEMS

Refraction: qualitative and empirical problems

SECTIONS
A. REFRACTION – THE BASIC PHENOMENON
B. QUALITATIVE PROBLEMS
C. QUANTITATIVE PROBLEMS USING TEMPLATE

A. REFRACTION – THE BASIC PHENOMENON

Exercises and problems on basic refraction phenomenon to be provided by instructor

B. REFRACTION - QUALITATIVE PROBLEMS

1. Spearing a fish from the river bank (Qualitative version)

A fish is under the water at the position shown, and a fisherman on the river bank hopes to spear the fish for dinner. One of the rays of light from the fish is shown going toward the surface.

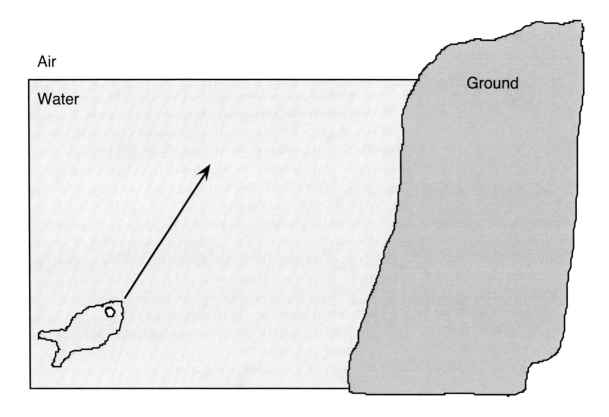

a) What happens to the ray as it passes through the interface between water and air? Draw in qualitatively the approximate path of the refracted ray which emerges into air.

b) Position the fisherman's eye so that he is able to see this particular refracted ray.

c) To the fisherman's eye, where does this ray *appear* to be originating from? Show on the diagram where the fish will *appear* to be located, according to the fisherman.

d) Will the fish appear to be *higher* or *lower* in the water than it really is?

e) If the fisherman throws the spear exactly in the direction that he sees the fish, what would happen? Would the spear *hit* the fish, or go *over* it, or *under* it?

f) So how should he aim if he wants to hit the fish? *At* the fish, *higher* than he sees it, or *lower* than he sees it?

g) Suppose the fisherman is holding the spear pointing directly in the direction he sees the fish. What would the fish see? Would it appear *to the fish* that the spear was aimed directly at her or not? Explain. So should the fish be afraid, very afraid?

C. REFRACTION: EMPIRICAL
– Quantitative problems using refraction template

Exercises and problems on ray tracing across an interface to be provided by instructor

Chapter 12

REFRACTION – quantitative

– a law for refraction

In this chapter we do a *quantitative* investigation, seeking a *law* for refraction, i.e. a relationship between the directions of the incident and refracted rays at an interface.

This investigation in search of a law will also serve as a wonderful example of science-in-the-making. Posing the question, planning an investigation, dong experiments, taking data, seeking relationships in the data, considering various possibilities or hypotheses, analyzing the data in terms of each, discovering a relationship that works, and finally proposing a *law* for refraction.

Then we use the law we have discovered to solve quantitative refraction problems, including the location of images formed by refraction.

PROBLEMS

12-1 INTRODUCTION

Our goal here will be to investigate refraction quantitatively. That is, we will send in light rays in specified directions, and measure what direction the refracted rays have. From this data we hope see if there is a relationship between the directions of the incident and refracted rays, and if so, propose this as a law.

Since we will have to specify ray directions quantitatively, we first consider how we can do this. We make no assumptions about how best to do this, but pose the question from scratch: what ways are there of specifying the *direction* of a ray?

12-2 SPECIFYING RAY DIRECTION

What ways are there of specifying the direction of a ray?

In order to be quantitative about refraction behavior, we need to be able to *specify the directions* of the incident and refracted rays.

The simplest refraction situation is when a light ray is 'head-on' to the surface. This ray is said to be 'normal' to the surface. Other rays will be offset from the normal, to a greater or lesser degree. How can we specify this? *We need to think of possible ways of specifying ray direction,* i.e. how much a ray is offset or angled. Can you think of two different possibilities?

One possibility is to state the ray's **angle** away from the normal. Another is to specify an **offset distance** from the normal, as a **semi-chord** of a reference circle. Both the angle and the offset distance (semi-chord length) are marked on the diagram alongside.

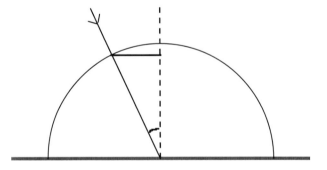

The instructor, using the examples provided, will discuss both the angle and offset (semi-chord) specification methods.

Trying the methods out

Try specifying the direction of the light ray in the diagram above, both as angle and as semichord. Do this by measurement on the diagram. That is, measure:

   i. The angle from the normal, and

   ii. The offset distance from the normal, i.e. the semi-chord of the reference circle. (You can also specify the radius of the circle, though it will turn out that semi-chord alone will be sufficient if we use the reference circle for both incident and refracted rays).

     • Angle value: . . . .

     • Offset distance (semi-chord length): . . . .

Note that the offset distance is for this chosen radius of reference circle. We will see later that it doesn't matter what size circle we choose, as long as the refracted ray is referred to the same one.

Discussion:

Note for the trig-savvy

In this book we don't assume you know trigonometry, but in case you do, here is a note of interest. It is actually the semi-chord *relative to the circle radius* that is the most useful quantity. That is, the semi-chord expressed as a fraction of the radius, rather than just the semi-chord alone. This 'ratio' way of specifying direction makes the measure independent of the size of the reference circle. Those who know trig will recognize the sine function :). But trig will not be necessary for our treatment to follow.

Using angles and semi-chords to specify ray directions in refraction

We can use either the angle method or the semichord method to specify ray directions in refraction. The refraction diagram below shows incident and refracted rays, and labels the angles and offsets for both.

REFRACTION: Diagram showing useful quantities

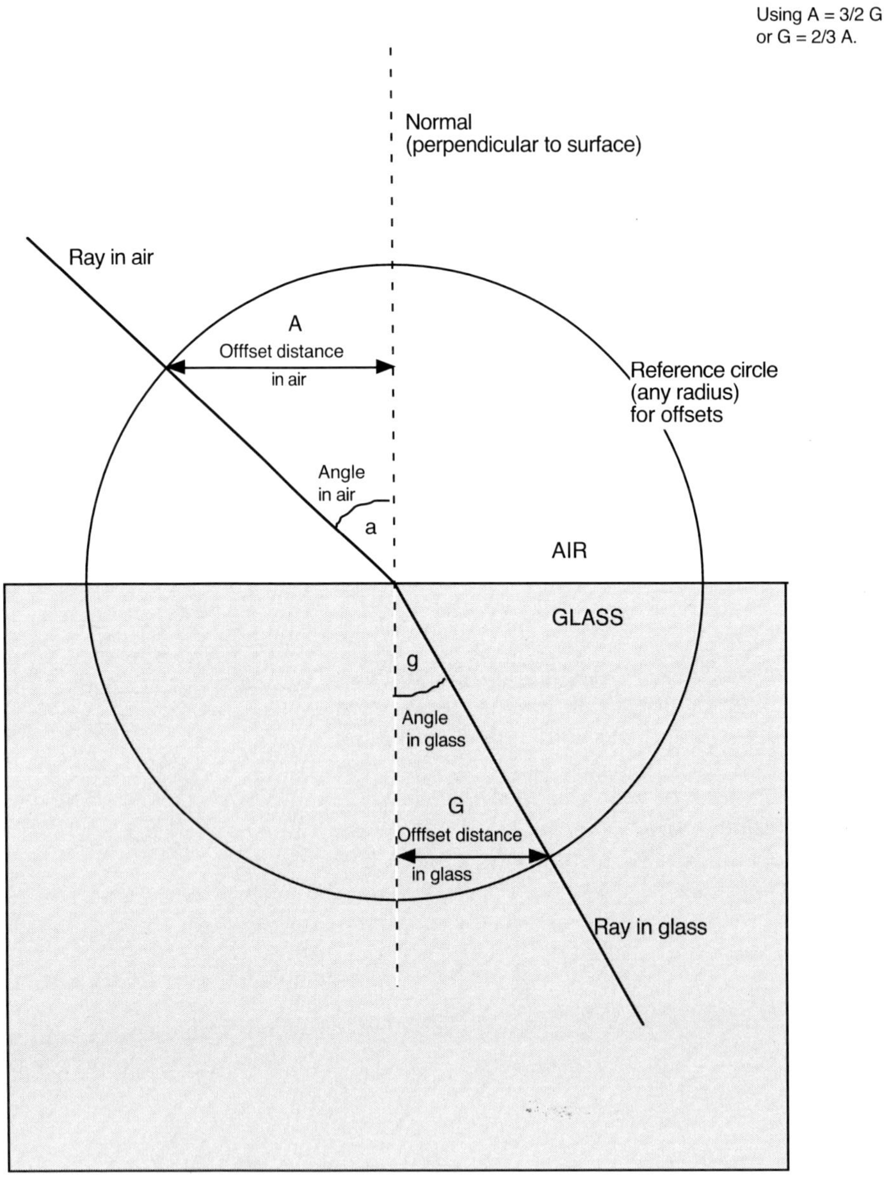

Using A = 3/2 G
or G = 2/3 A.

Normal
(perpendicular to surface)

Ray in air

A
Offffset distance
in air

Reference circle
(any radius)
for offsets

Angle
in air

a

AIR

GLASS

g

Angle
in glass

G
Offffset distance
in glass

Ray in glass

DGS FEB 2002

12-3 QUANTITATIVE INVESTIGATION OF REFRACTION
– seeking a law

1. Formulation of research question

We now want to investigate refraction quantitatively. What exactly should we aim to investigate about refraction? What will be out aim? Formulate your research question(s).

Designing your experiment

Plan and design a procedure to investigate your research question.

'Design' involves stating the goal, writing an initial plan of what you will do, how you will do it, and what data you will need to take. Be fairly specific, e.g. say what range of angles you intend to use, how many different readings, what you will measure, how you will use your data, what you will look for, etc. Plan also how you will record your data and design a table for your readings.

Doing the experiment

Carry out your experiment. Your 'raw' recording of ray directions will be by drawing them in accurately on paper beneath the glass block. Produce a set of diagrams showing the measured incident and refracted rays, for small, medium and large incident angles. (You can use more angles if you like, but at least three).

These recorded ray diagrams constitute your experimental data, and you will be able take measurements off them. Paste your diagrams into your lab notebook.

Pre-recorded refraction diagrams

We have also decided to supply some 'pre-recorded' refraction diagrams that the instructors have prepared in advance.

You have a choice here: you can either analyze your own ray data, or use the ray data that the instructors have prepared. We offer you prepared data for two reasons: it has been carefully done and set out clearly, which helps with the next stage. Also everyone in class can work with the same data and compare results. Prepared ray data for small, medium and large angles are on the separate sheets that follow. However, if your own diagrams are also clear, then of course you are welcome to use them.

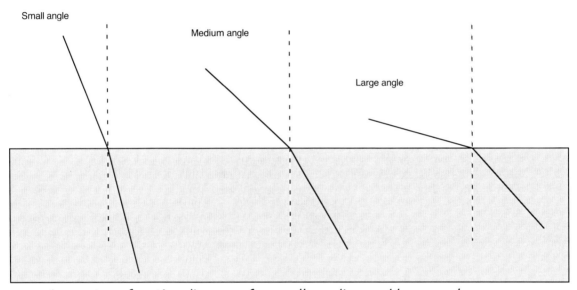

REFRACTION BETWEEN AIR AND GLASS

Accurate refraction diagrams for small, medium and large angles.

From your own diagrams or the pre-recorded diagrams, specify the ray directions (referred to the normal) by measuring both **angles** and **semi-chord distances.** Enter your data in the table below.

You have various directly **measured** quantities, the semi-chord lengths and the angles, as data. What other **calculated** quantities could you obtain from your data, that might be useful in looking for patterns, relationships, and laws? For example you could calculate sums, or differences, or ratios, or products etc. of the data that you have. Each of these will have a physical meaning.

Think about this. Don't look ahead at the table columns yet, but first think what calculated quantities **you** might like to try.

TABLE: Tabulate measured values of angles and semi-chords in air and glass.

Case	Incident angle	Refracted angle	Deviation angle	Incident semichord	Refracted semichord	Angle ratio	Semichord ratio	
Zero	Zero							
Small								
Medium								
Large								

A blank column has been added on the right for any other quantity you may want to calculate, and you can add columns if you wish.

Analysis of data, seeking and testing relationships, formulating a law

Analyze this tabulated experimental data, with the aim of seeking a rule for refraction. Is there any pattern you can find in the tabulated data? Is there perhaps something, or some combination, that remains constant no matter how the angles vary? Try various things.

Note that you should do this for both angle and semi-chord data, since we have no idea which of these might turn out more suitable.

Finally, can you find a rule relating the ray directions in air and glass? Which works, angles or semi-chords?

State your rule as a proposed 'law of refraction'.

Your Law of Refraction

The law you propose will constitute yet another 'powerful idea' for light behavior!

Reflections on knowledge – thinking about laws of nature

A law of nature has both beauty and power. Beauty in that it is a concise elegant synthesis of how nature behaves. Power in that we can use it to work out what will happen in a physical situation.

So it is with your law of refraction. For light, it is an elegant portrait and a powerful tool. It neatly rolls up all your observations about light bending into a single relationship which applies for any and all ray directions. And you can use it as a tool to work out exactly how a given light beam will bend when it strikes an interface.

"Ride a light ray and obey the law" – at the crossing!"

Note regarding trig formulation and 'Snells law'

Note that we have been able to obtain the law of refraction of light without assuming any knowledge of trigonometry. Instead we have used a reference circle and offset distances. It turns out that this procedure is in fact equivalent to using trig! So we have the advantage of being able to investigate refraction and find the correct refraction law, without having to learn trig first. Note that this is in fact the way Descartes formulated the law, centuries ago! We are walking in his footsteps. (It turns out that the author re-invented this method while developing this course, without knowing at the time that Descartes had beaten him to it by a long shot!).

When formulated using trig, the law of refraction is called Snell's law. The instructor will discuss it with you.

12-4 TESTING YOUR LAW OF REFRACTION

Put your proposed refraction law to the test, to see if it can successfully predict the outcome of a new experiment. Some testing experiments are given below.

1. Simple direct test

The most direct test of your law is: given a particular incoming ray, to predict the direction of the outgoing ray.

Lay a whiteboard on your bench. The instructor will lay a glass block on the whiteboard, and mark how a laser will be positioned later to send its beam toward the block.

Your task is to predict and draw in what direction the laser beam will have after refraction at the glass surface. When you are done, the instructor will bring the laser to try it out.

If your prediction is correct, you can have some confidence in the law and your ability to use it.

2. Other testing experiments

Provided by instructor as appropriate.

Do these prediction tests give you confidence that your refraction law accords with nature?

12-5 THE CONCEPT OF "REFRACTIVE INDEX"

Instructor will introduce and explain

EXAMPLE PROBLEMS

The instructor will provide example problems to illustrate the application of your law of refraction in different situations.

12-6 IMAGES DUE TO REFRACTION
– quantitative treatment

Image production by refraction

The instructor will introduce this section. By tracing two or more rays diverging from a point source, and undergoing refraction at an interface, you will be able to locate the image of the source.

12-7 REFRACTION AT CURVED SURFACES

The instructor will introduce this section. The same physics applies for refraction at an interface, whether the surface is plane or curved, just the geometry differs for sets of refracted rays, and thus the location of images will depend on the shape of the surface. Knowing the law of refraction, tracing rays one by one, and knowing how each bends at the interface, you are already able, in principle, to locate the image due to refraction at any curved interface! There is nothing you can't do by applying the basic law whenever a ray strikes an interface!

Note that one real life application of this is refraction by the cornea of the eye. The cornea is a curved interface between air on the outside and fluid on the inside of the eye, and refraction takes place as light passes from air into the eye.

PROBLEMS

Quantitative refraction problems using the law of refraction

---

1. Finding the refractive index of plastic

An experimenter is investigating how clear plastic refracts light. She aims a beam of light at a block of plastic, and finds that it is refracted as shown below. (The rays represent her experimental data and are drawn accurately).

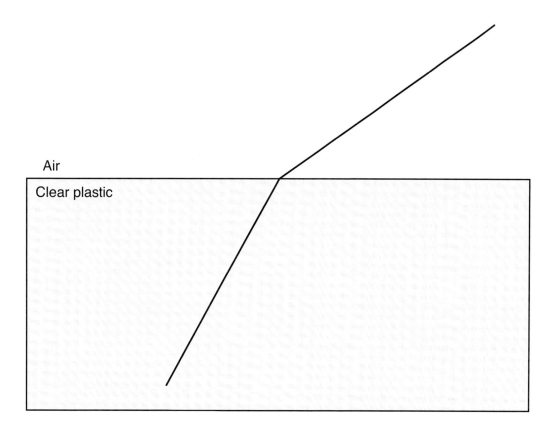

a) Find the ratio of semichords for refraction in plastic. Do this by the 'offset distance' method, constructing semichords and finding their ratio. This ratio of air to plastic semichords is called the "refractive index" of the plastic.

b) Hence construct the refracted path of any **other** ray B which is incident at a different angle. Choose your own ray B (or two of them if you like).

c) The refractive index of glass is known to be 1.5. State what this means, in terms of refraction semichords. Which will bend light more, glass or plastic? Explain your reasonin, referring to a sketch if you wish.

2. Spearing a fish from the river bank (quantitative)

A fish is under the water at the position shown, and a fisherman on the river bank hopes to spear the fish for dinner. One of the rays of light from the fish is shown going toward the surface.

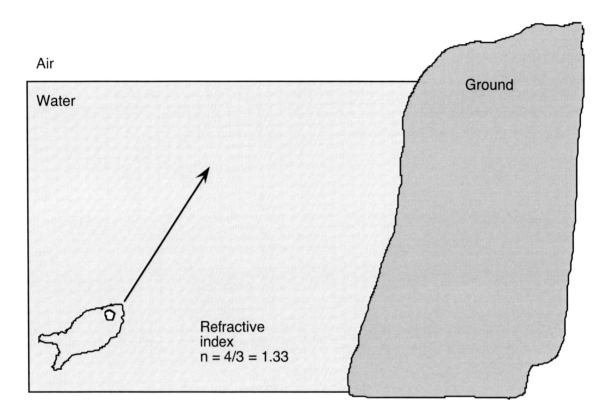

a) *Qualitative:* Sketch roughly what happens to the ray as it goes from water to air.   2

b) *Quantitative:* By accurate construction determine the exact path of the refracted ray in air. Use the offset ratio (semi-chord) method with a nice big circle. The refractive index of water with respect to air is 4/3 or 1.33, i.e. the ratio of air/water semi-chords is 4/3.   8

c) Position the fisherman's eye so that he is able to see this particular refracted ray.   2

d) Show on the diagram where the fish will *appear* to be located, according to the fisherman. 2

e) Will the fish appear to be *higher* or *lower* in the water than it really is?   2

f) If the fisherman throws the spear exactly in the direction that he sees the fish, what would happen? Would the spear *hit* the fish, or go *over* it, or *under* it?   2

g) So how should he aim if he wants to hit the fish? *At* the fish, *higher* than he sees it, or *lower* than he sees it?   2

h) Suppose the fisherman is holding the spear pointing directly in the direction he sees the fish. What would the fish see? Would it appear to the fish that the spear was aimed directly at her or not? . . . . .So should the fish be afraid, very afraid?   3

3. Refraction through a triangular prism

A beam of light is incident on a glass prism as shown. The refractive index of glass is 1.5.

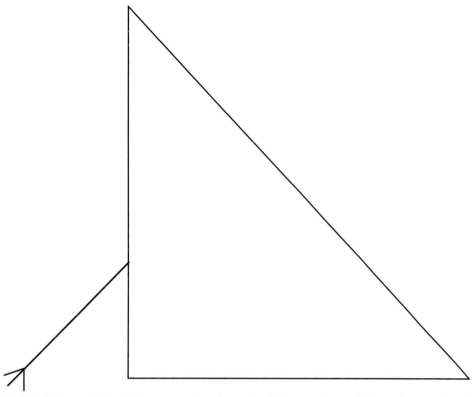

a) Sketch *qualitatively* the approximate path of the ray through the prism and out into the air again.

b) Find accurately the path of the ray that exits the prism. Do this by accurate ray construction, i.e. 'ride on a light ray and obey the law', using the offset (semi-chord) method for refracting light.

c) In addition to refraction, some *reflection* also occurs at the interfaces. Draw in accurately the *reflected* rays that occur.

CHAPTER 13

AFTERWORD
– THE CONTINUING STORY OF LIGHT

Through our observations, questions, and scientific investigations we have progressed quite far in our understanding of light and its properties. We have generated and tested concepts, modes and laws for the behavior of light under various circumstances.

But this is only a small part of the story of light! We have concentrated on 'geometrical optics', also called 'ray optics', i.e. those properties of light consistent with light traveling in straight lines and represented by rays. But there is more! Much more than we can hope to cover in this particular course.

Thus at this stage we would like to outline a 'big picture' of light and all its fascinating behavior and properties, so as to put our geometrical optics study into a broader perspective.

We start by recounting the progress we have made so far. The way we approached our study of light was to observe and investigate how light behaves in various situations, and hence formulate properties and principles for light. We aimed to develop a *model* for light, which embodied the aspects we discover.

On this scientific journey, we proceed by investigating light purposefully. Physics usually starts with the simplest situations, working toward the more complex. Thus we first explored how the light emitted from a small (point) source behaves. We listed the properties we discovered, along with the evidence for them, and devised a simple initial (ray) model for light. We also extended the ideas to larger (extended) light sources.

We also considered the role of the eye and how we 'see' things.

Some of the properties of light (like straight-line travel for example) have consequences which are interesting and significant in themselves, so we spent some time pursuing these (e.g. shadows, apertures etc). At the same time, we developed some of the mathematics and simulations that are useful in understanding light.

The next logical stage, though we were not able to do so, would be to consider *color*. We would suggest possible competing theories to account for color and its relation to white light; then test them. This also parallels what happened historically.

We turned to what happens when light interacts with matter, in different ways. This involves the phenomena of reflection, refraction, absorption and scattering. The knowledge produced turned out to be very useful, in that an understanding of this behavior can be used as a basis for controlling light and designing optical instruments.

As we proceeded, we developed and refine a model for light based on the properties discovered so far. This realm of study is called 'ray optics' or 'geometrical optics' and takes up most of our particular course on light.

All along our scientific journey, we *applied* the physics principles and models to *solving problems* in new situations. This also constitutes a real test of our understanding.

At this stage, we had some confidence in our understanding of the properties of light studied so far.

Note that we reached this understanding of light behavior without really tackling the thorny question of "what is light?" Historically, scientists put forward two ideas about this; namely that light might be like a stream of small particles, or light might be like a wave. There was considerable dispute about this between eminent scientists. We ask: would a particle model of light be consistent with all the properties we have observed so far? If so, the particle view would seem to work pretty well.

However the story is not so simple, and far from finished!

Historically, scientists extended the study of light and discovered more and different phenomena. The new properties of light revealed seemed to be at odds with those studied earlier, and even contradicted our earlier model! These aspects of light behavior caused scientists to favor a different model for light, to account for the new phenomena. This realm of light behavior is called 'wave optics'. It seemed that light could also behave as if it were a wave, and the wave model gained acceptance.

But nature had more surprises in store – yet *other* phenomena were discovered that could not be explained on the wave model! But the were consistent with light behaving like particles under certain circumstances. We will not be able to go into this, but can say something of the history. Scientists now had two models, addressing different aspects of light behavior. Eventually they devised a more comprehensive model for light, which could embrace the duality of these different aspects. The instructor will introduce and discuss this, in ending off the course with a glimpse of current thinking about light

NOTES